The Limits of Scientific Reasoning

The Limits of
Scientific Reasoning

David Faust

Foreword by Paul E. Meehl

University of Minnesota Press   MINNEAPOLIS

Published by the University of Minnesota Press,

2037 University Avenue Southeast, Minneapolis, MN 55414

Printed in the United States of America.

Library of Congress Cataloging in Publication Data

Faust, David, 1952-

   The limits of scientific reasoning.

   Bibliography: p.

   Includes index.

   1. Science—Philosophy. 2. Reasoning. 3. Judgment

(Logic) I. Title.

Q175.F269 1984   501   84-5172

ISBN 0-8166-1356-7

ISBN 0-8166-1359-1 (pbk.)

It may indeed prove to be far the most difficult and not the least important task for human reason rationally to comprehend its own limitations.

—F. A. Hayek

The counter-revolution of science (1955)

I am but a seeker after Truth. I claim to have found a way to it. I claim to be making a ceaseless effort to find it. But I admit that I have not yet found it. To find Truth completely is to realize oneself and one's destiny, *i.e.*, become perfect. I am painfully conscious of my imperfections, and therein lies all the strength I possess, because it is a rare thing for a man to know his own limitations.

—M. K. Gandhi (1921)

from *On Myself* (1972)

CONTENTS

Paul E. Meehl

I found this a fascinating, wide ranging, provocative and deeply questioning book which I should think will interest scholars in a variety of domains including scientifically oriented clinical practitioners, cognitive psychologists, psychometricians, statisticians, philosophers and historians of science, and last—but surely not least—academic, government, and foundation administrators concerned with policy questions in the funding of scientific research. I admit that seems a somewhat unlikely heterogeneous readership, but I think all these groups owe themselves an intellectual favor, and the taxpayer due consideration, by reading this book. It also has the effect of refurbishing my somewhat battered image of the "scientist-practitioner" as a training model in the field of clinical psychology, because Dr. Faust writes from the standpoint of a scientifically sophisticated, theoretically oriented *seasoned clinical practitioner*. One may ask whether the author's clinical experience was strictly necessary for his being able to write a book such as this, and I, of course, am not prepared to prove any such counterfactual thesis. But my conversations with him (plus, of course, the point of take off for his reasoning in the book itself, see below) lead me to think that this part of his life experience played a major role, and in any case the empirical fact is that it *was* a practicing clinician interested in cognitive processes rather than a pure academic theoretician who did produce the book.

The chief reason (other than my interest in matters methodological) that I was invited by Dr. Faust and the University Press to write an introduction to his book is that the starting point for his methodological reflections is the controversy concerning the relative efficiency of clinical and statistical prediction. This controversy — which goes back over 40 years of the social-science literature (e.g., Sarbin, Lundberg) and was pushed into prominence, leading to a large number of research studies, by my *Clinical versus Statistical Prediction* (1954) — has suffered the fate of so many controversies in psychology and other social sciences, to wit, that one group of persons considers it to have been settled, a second group continues to resist the massive empirical evidence and think it still controversial, and a sizable number of people don't even seem to know that an issue exists, or that it is of any practical importance. It is safe to say, as Dr. Faust summarizes early on in the book, that the mass and qualitative variety of investigations of the predictive efficiency of subjective impressionistic human judgment, such as that exercised by the individual clinician or case conference or psychiatric team, versus that of even a crude non-optimized mechanical prediction function (equation, monograph, actuarial table) is about as clearly decided in favor of the latter predictive mode as we can ever expect to get in the social sciences. *I am unaware of any other controversial matter in psychology for which the evidence is now so massive and almost 100% consistent in pointing in the same direction.* Psychoclinicians' arguments by analogy from traditional medicine are badly blunted by the finding that diagnosticians in "organic" specialties (e.g., radiology, pathology, internal medicine) show similar cognitive deficiencies in interjudge reliability, weight inconsistency, inferiority to more "objective" data-combination and inference-making methods. That this body of data has had so little effect upon clinical practice reflects on the scientific mental habits of practitioners and the defects of training programs. It is, alas, not unique because it can be paralleled with other examples, such as the continued reliance upon costly skill-demanding projective tests which have been repeatedly shown to have low *absolute* validity and negligible *incremental* validity for purposes of making practical decisions that matter to the welfare of the patient and use of the taxpayer's dollar.

Starting from this somewhat unlikely jump-off point, Dr. Faust generalizes the finding concerning the inefficiency of clinicians' inferential habits and uses of evidence in what I found initially to be a somewhat disconcerting way. Those of us clinicians who attempt to think scientifically and, therefore, pay attention to the body of empirical data bearing on the clinician's cognitive functioning have often—especially if we have a personal commitment to such noble traditions as, in my case, psychoanalysis—analogized processes of complex clinical reconstruction, "the integrated picture of the individual patient's structure and dynamics," to theory construction of the nomothetic sort. That is, while it may be plausible to say that a simple unweighted arithmetical sum of certain test scores or demographic or life-history facts is the most efficient way to forecast subsequent delinquency or survival in engineering school, we have the idea that we do not expect any such mechanical, automated, straightforward "clerical" process to be capable of fathoming the patient's mind in more theoretically interesting respects. So we make a special place for this in our thinking, and I confess that I continue to do this, especially with regard to psychoanalytic interpretation of a moderately complicated dream. But I am aware that these special clinician's hold-outs against objectification of the inferential process might merely be sort of wishful-thinking "last-stand" positions which the actuarial enemy has not as yet approached with grenade and flame thrower! Logicians have pointed out repeatedly that "theories" in the life sciences, and even in certain portions of the inorganic disciplines (e.g., historical geology), are not cosmological theories of completely general nature, such as Maxwell's equations or Quantum Mechanics which, as Professor Feyerabend puts it, "claim to say something about everything that happens." So that what appear at first blush to be general scientific theories, thought of as nomothetic rather than idiographic (e.g., the theory of evolution, the theory of continental drift, or Freud's 1895 theory about the specific life-history origins of hysteria versus obsessional neurosis) are apparently nomothetic but when examined closely turn out to be idiographic in character. However soothing these metatheoretical thoughts may be to the hard-pressed psychodynamic clinician (like myself), what does Dr. Faust proceed to do, taking off from the shocking findings regarding the poor cognitive efficiency of clinical inference? Instead of containing that distressing generalization by analogizing *other*, more complicated sorts of clinical inference to scientific theories (with the associated consoling thought that in that case they are in pretty good shape although they might need a little tightening up here and there), he turns the argument on its head, generalizes the statement about low inferential competence, and carries the battle into the field of scientific theory construction itself. This is a clever and daring move on his part which, even if the reader ends up rejecting it, can hardly fail to be of interest to a thoughtful person. He challenges as unproved, as not antecedently very probable, the widely held notion— hardly questioned by most of us, including a somewhat hard-boiled cynic like myself—that run of mine scientific thinking, even by reasonably competent practitioners of the scientist's craft, is anywhere optimal in strategy and tactics. He bolsters this criticism— which, after all, could almost stand on its own feet, once raised— by evidence from recent research by cognitive psychologists into the problem-solving and decision processes, as well as nonexperimental data taken from the history of science. So that analogously to saying after surveying the literature on clinical prediction, "Well, we may be surprised and disappointed by these studies, but, after all, why should we have *expected* the clinician or the clinical team to be a highly efficient information processor and inference maker?" he proceeds by saying, "What grounds are there, theoretical or empirical, for supposing that most scientific reasoning, whether about theories or about the design and interpretation of experiments, is of high quality?"

To enter wholeheartedly and open-mindedly into the ramifications of this development in Dr. Faust's thought, I found it necessary to shed, or at least momentarily set aside, some of the residues of the logical positivism that I learned almost half a century ago. For example, as an undergraduate in the late 1930s I heard from lecturers, and saw repeatedly in articles or textbooks on philosophy of science, that the logician and philosopher of science is not so arrogant as to try to *prescribe* for the scientist how he should go about his business, but that the aim of the philosophy of science discipline is to give a rational reconstruction of scientific knowledge, or even, in the extremely sanitized form held by Carnap, narrowly defined as giving the syntax (and, later, the semantics) of the "language" of science. Now anyone, whether scientist (perhaps likely) or philosopher (hardly likely today), who still holds to that hypermodest view of the philosopher's enterprise will probably be made a little nervous by some of Dr. Faust's text, and especially some of the constructive suggestions for improving scientific reasoning that appear later in the book.

I remember as a student being somewhat puzzled by this becoming modesty of the philosopher in his insistence that he was only describing (or analyzing, or at most "reconstructing") what the scientist does and certainly not laying down any methodological rules or in any way trying to pass judgment upon the scientist. That is, the enterprise is purely *descriptive* (although in a funny sense of that term, being a mixture of empirical and formal subject matter) but not at all *prescriptive*. Now one knows that there are no algorithms in inductive logic that are of such a sort as to be immediately applicable to a quantification of inductive scientific reasoning. Although, of course, there are some for specialized subfields of inductive inference such as Bayes' Theorem, and Fisher's maximum likelihood criterion for choice of a statistical estimator. Nevertheless, it would seem that if we allow ourselves the jurisprudent's distinction between "principles" and "rules," between some admittedly vague guidelines or rules of thumb that one can sometimes fruitfully violate (but only at a certain statistical peril), and some strict constraints that it is irrational, incoherent, or maybe downright senseless to transcend, then it's hard to believe that the result of the philosopher's reconstruction of scientific knowledge would not even contain any such broad principles. After all, philosophers of science and logicians distinguish their enterprise from that of historians, and the latter are equally adamant about maintaining the distinction. All one needs to do is to pick up a book or article written by a historian of science, and then one written by a logician or philosopher of science, to see that while the clarity and sharpness of the distinction has perhaps been exaggerated (for complicated reasons as I would hold), there is certainly a striking difference between the two in respect to both goals and methods. But if philosophy of science as "reconstruction" differs importantly from pure non-philosophical history of science, it is hard to see what *is* its true subject matter, if there are no principles of rationality or of conceptual clarity, no prescriptions about how to frame definitions and how to detect flaws in suggested concept formations, no way to explain or justify the confirmation of theories, and the like. Faced with this paradox, we may conclude that Carnap and Co. were excessively modest, perhaps because they were great admirers of the scientific edifice and antagonistic to traditional philosophy (being greatly embarrassed by such philosphical bloopers as Hegel's trying to dictate to scientists on metaphysical armchair grounds how many planets there could be, or thinkers deciding philosophically about determinism rather than looking to experimental physics). We conclude tentatively that they should have said forthrightly that to the extent that any kind of rational reconstruction of scientific knowledge can be successfully carried through, it will, when surveyed over a set of advanced scientific disciplines, permit some generalizations not only about *rules* of the game (e.g., one cannot tendentiously select which data to include as evidence, literally ignoring the rest), but also about strategies that *tend* to have good scientific payoff (not "rules" but hints, helps, guidelines, "good advice in general") and, furthermore, that this latter conclusion will not come from the logician's armchair but from reflections upon the history of science itself. But this means that the philosopher of science will, if the discipline succeeds in getting anywhere with its *own* aims, be in a position to criticize the scientist and to give advice. At a less sophisticated level this almost seems obvious, because surely the professional logician would have no hesitation in pointing out a formal fallacy committed by a scientist in a theoretical article; scientists criticize one another for a material fallacy such as the fallacy of Neglected Aspect; or for forgetting the asymmetry between corroboration and falsification; or for ignoring the fact that the alleged dozen replications of a certain experimental result all come from one of two laboratories and the director of the second was a Ph.D. trained at the first; or a statistician points out that a certain estimator is biased, or reminds the scientist that one pays a price for correcting a bias in the form of increased sampling error variance. All these critical, prescriptive, normative comments being common and accepted scientific practice, it seems strange to say that the results of such a scholarly enterprise as philosophy-cum-history of science cannot include helpful advice and at times, at least, a possibility of powerful negative criticism.

We were also taught in my youth that philosophy of science was not empirical and that it was necessary always to be crystal clear in one's thinking about whether one was operating in Reichenbach's "context of discovery" or "context of justification." Looking back today, I find it somewhat strange that we received this so uncritically. I remember a conversation with the late Grover Maxwell in which I was dragging my feet in connection with some aspects of his and Donald Campbell's epistemology, on the grounds that we want epistemology not to be conflated, however slightly, with empirical psychology and sociology of knowledge. Maxwell cured me of that absurd statement by a single question, which I have subsequently come to think of as "Maxwell's Thunderbolt" (which should be added to better known destructive tools such as Occam's Razor and Hume's Guillotine). "Well, Meehl, what epistemological or methodological statements can you derive *from logic alone?*" One immediately sees that such standard old Vienna "rules of the scientific game" as intersubjective testability are rooted in social and biological generalizations about human brains being connected to sense organs, people having overlapping perceptual fields, most of us who are sane and have corrected vision being able to see pretty much the same things under carefully specified conditions, etc. One thinks of B. F. Skinner's famous answer to Boring as to the intersubjectivity criterion, that whether Robinson Crusoe has developed a good science of ornithology or botany on his desert island depends not upon his being joined by his man Friday (with whom he can have "intersubjective agreement" on some sentences) but upon whether the sentences that Robinson Crusoe speaks to himself enable him to predict what happens and gain scientific control of his subject matter! The obvious point is that in the life sciences there are certain commonsensical theories about the day-to-day persistence of macroscopic bodies, about the limitations of human sense organs, the fallibility of human memory (leading us to record observations at the time) and the like, which do not normally give rise to any deep questions in cognitive psychology but which are presupposed in any epistemological discussion and *a fortiori*, any discussion of that subdivision of applied epistemology we call "methodology of science." It is a result of physiological facts about the human animal, for instance, that we erect physics and chemistry and astronomy upon the deliverances of the eye more than the ear and of the distance receptors more than smell and touch, it being an empirical fact (as N. R. Campbell pointed out in 1920) that essential unanimity of judgments by normal, trained, careful, and honest persons can be reached for judgments of spatial coincidence or betweenness, temporal succession, and number.

I may say here, and I am not sure whether this is a point that Dr. Faust and I would be in agreement about, that I see a certain danger in the current emphasis upon cognitive psychology in the social sciences as bearing upon methodology. As I read the record, the advances in the other sciences where human perceptual distortions or individual differences played an important role did not occur primarily by making corrections for them (despite the famous origin of experimental psychology in the "personal equation" discovered as a result of the disagreement in star transit observations by Maskelyne and Kinnebrook) or, to any great extent, by the perceptual training of the observers (although that is sometimes unavoidable as, say, in pathology), but primarily by replacing the human sense organ and especially the human memory with some other device. Today we see replacement of a human making a measurement by an electronic device making the measurement and what the human eye has delivered to it is a computer printout. I see a certain obscurantist danger lurking in the background for psychologists and—perhaps more for sociologists and anthropologists—in current enthusiasm for the theory-dependence and observer-infection emphasized by Kuhn, Feyerabend, Hanson and Co., because I do not see the history of the more developed sciences reflecting a positive *building in* of the human element, or even *correcting for it*, so much as a systematic *elimination of it* at every stage where it tends appreciably to determine the protocol results.

That brings me to the only other place where I would like to have had more chance to debate with Dr. Faust because I think we disagree somewhat. He accepts happily the current emphasis upon the theory infection of observations, and I do not believe that accepting it to that extent is a necessary part of his thesis, although he might be able to convince me otherwise by more extended discussion. It seems to me that whereas our theories (and I include here such low-order "theories" as predilections for how to slice up the flux of sensory experience into appropriate chunks and the discernment of meaningful recurrences of "more of the same"), whether implicit or explicit, do motivate our bothering to observe at all, will help determine which features of the isolated situation we control (eliminate, hold fixed, or ourselves manipulate to various levels), which events under those controlled conditions we select to pay attention to, which aspects of such selected events we pay attention to, how we break up the properties of the aspects for recording purposes and, if they exist in degrees, what metric we choose, and finally, what kind of mathematical manipulation of these data we engage in—all of these are partly theory-determined and some of them may be completely theory-determined. But it remains the case that *given* the above list having been settled for a particular research study, then what the *results* are, which numbers or which qualitative predicates appear in the protocol, must not be theory-determined, because if it were the argument would be viciously circular and we would not be empiricists at all. The short way of saying that is that if the theory, in addition to determining all the other things I have mentioned, determined the outcome, the *numbers*, or the *predicates* in the protocols themselves, then we could save ourselves a lot of time by not going into the lab or the field or the clinic, since we already know the answer. And, of course, this important point which still defines the empiricist (whatever his philosophy of science may be) must not be conflated with the important point (neglected by most Vienna positivists and by some tough-minded psychologists even today) that there are occasions by which the candidate protocol is properly excluded from the corpus on theoretical grounds. But we must be careful here: The exclusion of a candidate protocol from the corpus because a strongly corroborated theory forbids it is *not* the same as the theoretical ascertainment of the content of a candidate protocol, this latter purporting to describe observations we made in the laboratory. The widely heard expression, "Theory determines in part how you interpret what you see" badly needs parsing, because it has several interpretations, some of which are correct, some of which are misleading, and others of which are just plain wrong. So that even Otto Neurath's infamous business that caused Bertrand Russell so much distress, that we will admit a candidate protocol if we find we can "fit it in [*eingliedern*]," means that *given* the protocol offered by the investigator, we decide whether to "include it in," meanwhile hoping—but not requiring—that we will be able to explain its occurrence if it is excluded. Neurath surely did not mean here that the observer chooses to record the number "17° C" or the predicate "green hue" either whimsically or on the basis of belief in some theory. What the protocol of an honest reporter (right or wrong, but still honest) *says* he saw or heard must be distinguished from the decision of the scientific community, which may include this observer himself, whether to receive it into the corpus. Whether these worries of mine about the theory infection or theory dependence of observations would be agreeable to Dr. Faust I do not know; but it does not seem to me that any of the main points in the book would require that he disagree with them in the interest of consistency.

Dr. Faust's analysis of sources of judgment errors into analytic categories should be helpful both to psychologists researching cognitive

processes in complex situations and to clinical teachers and supervisors trying to teach and train clinicians to reason more effectively.
One hopes that clinical readers will take seriously his challenging of
what is probably the clinician's most common defense mechanism,
faced with such a large body of discouraging data about his or her
own performance, which is to say that the unique superiority of the
clinical brain is manifested in complex cognitive tasks even if it doesn't
show up in simple ones; or, if the basic variables are fairly simple and
few in number, in the complexity of their optimal combination in a
configural function, a plausible guess about results of future research
that I permitted myself in 1954 and attempted unsuccessfully to
exemplify in MMPI profile interpretation.

Dr. Faust's hard-hitting discussion of the rather unsatisfactory
character of literature reviews I hope will be taken with the serious-
ness it deserves. My prediction is that it will not, because in the behavioral sciences (especially in the soft areas like clinical, personality,
social, and counseling psychology) the relatively weak state of theories
and the modest reliability and validity of measuring instruments
makes it difficult—in many sectors impossible—to generate numerical point predictions, or narrow-range predictions, or even predictions
of the mathematical function form with parameters to be filled in.
Hence theory testing in these fields is frequently restricted to performing of directional statistical significance tests. The almost exclusive reliance upon establishment of a non-null difference between two
groups, or a non-zero correlation between two variables, makes the
usual literature summary well nigh uninterpretable by a sophisticated
reader. I realize this is a strong statement—stronger than Dr. Faust
makes, although he comes mighty close to it—and I am preparing a
scholarly paper that will I hope prove the point. When one is forced
to rely upon scanning a heap of significance tests, some of which pan
out and others not, there are six nuisance factors working that are
variable, countervailing, of *unknown* but *non-negligible magnitude.*
These six factors are: Lack of strict deducibility of predictions from
the theory, problematic character of unavoidable auxiliary hypotheses, inadequate statistical power, influence of the ubiquitous "crud
factor" (in social science everything is correlated with everything and
the null hypothesis taken literally is always false), bias in submission
of reports in favor of those that come out significant, and editorial
preference for acceptance of significant results because of Fisher's

point that the null-hypothesis can be refuted but cannot be proved. The first three of these tend to work against successful corroboration of a good theory; the latter three work in favor of pseudo-corroboration of theories having negligible verisimilitude. Since one does not know in a particular domain the relative magnitudes of these six opposed forces, it is usually not possible to interpret the box score of a heap of significance tests in the soft areas of psychology.

The author suggests that one decision aid that might extend the scientist's capacities to make judgments would be to use actuarial methods to evaluate theories. This suggestion will come as a shock to those conventional psychologists who suffer the delusion that this is already being done, a statistical significance test being (in its way) an "actuarial" method. But that, of course, is not at all what Faust is talking about here. He is talking about a more souped-up kind of actuarial thinking in which preferred theories are found actuarially, that is, by the study of the history of science in various domains, to possess certain signs of parameters. (I hope I do not misunderstand his intentions here, since the discussion of this point in the book is quite compact.) This notion revives an idea put forth almost a half-century ago by Hans Reichenbach in his *Experience and Prediction* (1938). Reichenbach attempted to defend his "identity theory" of probability, that all probabilities are really relative frequencies, including those "logical" probabilities represented in the concept of degree of confirmation or degree of evidentiary support, considered by Keynes and Carnap (and by most scholars today) to be fundamentally different from the notion of relative frequency. When we find ourselves critically examining the internal structure of a theory in relationship to its evidentiary support, it certainly does not *look* as though we were doing any kind of frequency counting. But Reichenbach's idea—which he did not spell out in sufficient detail to know whether it makes sense—was that various features of that relationship, and various internal properties of the theories, are empirically correlated with long-term success ratios. I daresay most philosophers and historians of science will drag their feet about Dr. Faust's suggestion that we could apply actuarial procedures to the life history of a major theory over a period of time, but once you get used to the idea it seems reasonable and probably do-able. I have often thought that even competent historians of science, or philosophers of science relying heavily upon history of science arguments in support of their methodological proposals, are not sufficiently conscious of the extent to which such history-based arguments are inherently *statistical* in nature, because they involve a sampling of episodes. Controversies about falsifiability and theory-based exclusion, ad hocness, and incommensurability, paradigm shifts which continue to rage among philosophers of science are replete with historical examples showing either that somebody did stick profitably to a degenerating research program, or abandoned it wisely, that somebody accepted what appeared to be *prima facie* a definitive falsifier and somebody else (as it turns out wisely, after the fact) did not, and the like; one has the impression that the authors do not fully realize that this kind of debate is difficult to move unless some sort of sensible claim can be made about the manner in which the theories, investigators, and episodes were selected. You can prove almost anything by historical examples if allowed to choose them tendentiously.

The most shocking suggestion of all, and one which, in addition to its novelty, is deeply threatening to the scientist's self-image, is the notion of using a computer to invent theories. While there is no algorithm for theory construction comparable to those for testing the validity of a complex statement in the propositional calculus (which computers have been able to do for many years) now that we have reached the stage where computers can play grand-master-level chess — and there are recorded instances of grand masters being defeated by computers — it is surely rash for the scientist to say that no kind of algorithm for concocting theories could ever be built. That's a fascinating possibility, and one can speculate about kinds of meta principles to embody in the program. Example: We say there are four basic kinds of entities in the world (structures, events, states, dispositions), and there is a quite limited number of antecedently plausible ways in which these four kinds of entities can be put together in a theoretical postulate set. We set the computer to work trying them out vis-a-vis a fact collection. This is music of the future, but it is to be hoped that readers will not dismiss Faust's thoughts along these lines.

I would have liked to see some discussion of the extent to which garden-variety mistakes of reasoning can be reduced in frequency by systematic "inoculation" instruction at the graduate level and to what degree such instruction has to be tied to a particular subject-matter domain. For example, reports of the brighter, more scientifically oriented clinical psychology students about their internship experiences convince me that many clinical supervisors functioning in those clinical installations do not grasp the importance of Bayes' Theorem in clinical decision making and information collecting, despite the fact that a clear, vigorous, and widely cited presentation of this methodological point was made by Albert Rosen and myself 30 years ago, and had been fairly clearly made by psychologists in the area of personnel psychology a half-century ago! It turns out that we need not invoke any interesting and complicated explanations of this social phenomenon, because the fact is that the reason clinicians don't think that way is that nobody ever told them about it or assigned it in their reading when they were in graduate school. As an undergraduate I was exposed to one of the "greats" in applied psychology, Professor Donald G. Paterson, in a course in differential psychology which relied on the textbook for certain important but not thrilling factual matters, so that Paterson could devote class time to critical examination of empirical studies that purported to show something that they didn't show because they were methodologically defective. Taking studies apart piece by piece and bone by bone, with the constant reappearance of certain bloopers (e.g., failure to control for selective migration, spurious index correlation, subtle contamination of ratings, insufficient statistical power, failing to consider effects on correlation of restricted range, use of an inappropriate descriptive statistic, picking out a few significant differences from a much larger batch of comparisons initially made, the assumption that a variable like social class is always an input causal variable and never an output variable itself influenced by genetic factors of intelligence or temperament)—this reiteration of certain basic methodological warnings in a diversity of empirical contexts is analogous to working through in psychoanalysis. I am willing to conjecture for falsification that there is no substitute for this kind of pedagogy if you want to teach psychology students to think straight about what research proves or doesn't prove. When I served for five years on the American Board of Professional Psychology, examining clinicians who had received their doctorates from accredited schools, I was aghast at how many of them had never had such course and (as a result?) had a naive approach to what was known and what had been proved. I would be curious to know whether Dr. Faust thinks that this is an over-optimistic view of the impact of formal classroom instruction. Undergraduate courses in general logic might be thought beneficial in that respect, and I do not have any hard data as to whether they are, but my anecdotal impression is that, for some reason not clear to me, such courses often do not "take" to the degree that one might hope. Is this because of their generality? After all, the various formal and material fallacies found in logic textbooks are usually exemplified so as to cover a wide range of content which would, if it fails pedagogically, look like my conjecture is falsified.

To conclude on an optimistic note, Dr. Faust and I and the reviewer and publisher must have faith that the right sort of verbal instruction can sometimes improve scholars' problem-solving practices, else why have we thought it good to publish this book?

PREFACE

The central thesis of this book is simple – that human judgment is far more limited than we have typically believed and that all individuals, scientists included, have a surprisingly restricted capacity to manage or interpret complex information. This thesis implies that scientists do not perform many judgment tasks nearly as well as we have assumed that they do and that there are many judgment tasks the scientist is expected to perform but cannot because of restrictions in cognitive capacity. If this is the case, it may be necessary to revise many past and contemporary descriptions of scientific activities and to alter the content and thrust of many prescriptive programs for science.

Given these implications, objections to the central thesis are to be expected, although I have been somewhat amazed by the vigor with which these objections have been voiced. Typical arguments – "The group overcomes the limitations of individual scientists" or "Scientific method *ensures* protection from cognitive limitations" – are put forth as self-evident, with little critical attempt made to consider the substantive issues raised by the judgment literature.

I do not pretend to understand the source of these almost reflexive negative reactions, but I will venture a few guesses. At least on the surface, the argument that scientists labor under stringent cognitive limitations may appear counter to sound belief. If scientists were so limited, how could they have accomplished what they have accomplished? Although it is possible that negative reactions stemming from this basis may be valid because the primary thesis is not (valid), it is also true that the common sense of today is commonly wrong tomorrow. I would be troubled if my arguments were rejected solely because they were abrasive to commonsense notions.

Negative reactions may also stem from the view that the central thesis is derisive of human beings and their capabilities. As a colleague once remarked, "If someone asks for constructive criticism, tell them something good, because they don't really want to hear anything bad." In a way, the "news" about human cognitive capacity *is bad*. Even worse, the position put forth here regarding human cognitive limitations is accompanied by the message that machines or mechanical procedures can often surpass human performance and will increasingly outstrip our capabilities over time. We were once jarred from the belief that the heavenly bodies spin around us, and the belief that we are at the center of the intellectual universe is now being threatened. In an age of dehumanization, a time in which more and more tasks are taken out of the hands of individuals and transferred to cold machinery, we are ambivalent about further moves in this direction. Those whose work is thinking are just beginning to experience the displacement that physical laborers suffered during the industrial revolution. For we are certainly entering into, or are in, the information revolution, and there can be little doubt that thinking machines will replace thinking persons in many realms.

While acknowledging the many benefits of this revolution, there is no denying its tragic side. I deeply regret that this book may reflect, or even contribute in some small way to, the tragic outgrowths of our progress. Spirited resistance is not without justification. This is not to say that an awareness of limitations, overall, is bad, or that the development of external needs for overcoming these limitations, overall, cannot produce many more gains than losses.

Other possible reasons for what I consider premature negative reactions to the central thesis include the potentially misleading nature of introspective experience, or the fact that the judgment literature directly challenges many long held and widely accepted beliefs. There is no need to dwell upon these matters here; many possible counterpoints to the central thesis are discussed in detail in the text. I simply wish to state my belief that the judgment literature makes a powerful case for the central thesis and that the study of scientists' cognitive limitations, and means for transcending these limitations, are issues of first priority. I am convinced that the cognitive limitations of scientists will be increasingly recognized over time and that such recognition will have a profound impact upon our view of science. The issues are too important to dismiss in a perfunctory manner, and it is my hope that the arguments put forth in this book will not be so treated.

Thanks are in order to numerous individuals who contributed to this work. I am grateful to both Barry Singer and Michael Mahoney, who reviewed a preliminary draft of this book and provided insightful comments and germane criticisms. My editor, Beverly Kaemmer, exercised both patience and skill that always proved helpful. Any typist who can wade through my initial drafts deserves the Purple Heart. I gratefully bestow this honorary award to my two typists, Cheryl Yano and Sue Rosenfield, for their tolerance and courage in tackling these materials. I also wish to give special thanks to my wife, Pam Stevens Faust, who, while I write this, is once more kept waiting late, and who only complained about my long hours when I fully deserved or needed it. Finally, at a time when I had met with nothing but negative reactions to this work, Paul Meehl provided the first source of encouragement and moral support. I owe him both an intellectual and emotional debt that I doubt he fully realizes and that I will always remember and appreciate.

David Faust

Providence, Rhode Island

The Limits of Scientific Reasoning

CHAPTER 1

Basic Assumptions and the Psychology of Science

Scientists, along with all other individuals, evidence cognitive limitations that lead to frequent judgment error and that set surprisingly harsh restrictions on the capacity to manage complex information and to make decisions. In the case of the scientist, this limitation extends to the judgments required by most prescriptive programs for science. This thesis implies that many descriptions of science must be inaccurate because they assume or report activities that go beyond the bounds of the scientist's cognitive capacities. In consequence, many prescriptions for science must be revised in some manner so that the scientist can meet their judgment demands.

Research on human judgment is used to support this central thesis, but it is also dependent, to varying degrees, on a series of underlying assumptions: that scientific knowledge does not come prepackaged but is a product of human cognitive activity; that the information relevant to the great majority of scientific judgments is complex; that individual human cognition makes a central, unique, and necessary contribution to scientific constructions of the world; and that scientific results do and should contribute to the philosophy of science. These assumptions, with the possible exception of the first, are controversial. Within this first chapter, I discuss them more fully and describe a potential role for the psychology of science.

COMPLEXITY OF INFORMATION RELEVANT TO SCIENTIFIC JUDGMENTS

The Place of Human Cognition in Science

Thinkers have long struggled with the role that human cognition does and should play in the acquisition of scientific knowledge. Views about the role of cognition, including its assumed merits and liabilities, have changed as our views of science and human reason have changed. We have gradually moved away from the view that science uncovers knowledge that directly mirrors external reality, and toward the belief that science produces constructions or interpretations of the world. Although the Greeks proffered logic as the vehicle by which the truth status of empirical statements could be unequivocally determined, we have come to realize the limitations of such an approach. We have abandoned the belief that tests of logic, or any other such definitive method, could be applied to all scientific statements, and we have slowly come to accept the view that even factual statements cannot be definitively evaluated for their truth status. We no longer believe that pure reason applied to pure facts provides pure knowledge. We no longer even believe that pure facts are possible, nor that there is a logic that can be used to decipher what the facts might mean.

This shift in viewpoint has lead to a reconceptualization of the role of human cognition in science. Human cognition was previously seen, at best, as a receptacle for knowledge. Any constructive interpretation was regarded as undesirable, as contamination of the data. From this framework, cognitive construction was considered antithetical to true scientific knowledge, a factor that should, and potentially could, be eliminated. We now realize that human cognition makes a unique and necessary contribution to scientific knowledge and that, without cognitive construction, there would be no scientific knowledge. To admit this does not necessitate the retreat into fictionalism that some have chosen—the view that scientific knowledge reflects nothing but the inner workings and imagination of minds, and that the external world contributes nothing to what is known. Rather, one acknowledges the contribution of the knower to the known. It is an awareness that human thinking must be actively applied in constructing a scientific picture of the world, that obtaining direct reflections of the external world is not only impossible but would likely render nothing but a buzzing mass of confusion. It is neither possible *nor desirable* to eliminate cognitive construction from science.

Recognition of the central role played by human cognition in constructing scientific knowledge not only implies that human cognitive activity must be considered in *any* comprehensive description of, or prescription for, science, but also leads to a critical problem shift. Rather than attempting to eliminate the contribution of the knower to the known, the aim becomes one of understanding how and how well the knower extracts information from the world, and of optimizing the effectiveness of his or her efforts to do so.

The Complexity Inherent in Scientific Judgments

Before examining the scientist as problem solver, we must consider the characteristic features of scientific problems or problem-solving tasks. There would be limited value in studying the role of human cognition in science if scientific judgment tasks were simple and easily solved. But this obviously is not the case. The importance of human cognition, and particularly of cognitive limitations, stems directly from the difficulty inherent in the judgment tasks confronting the scientist. A complex set of information and considerations is relevant to virtually all forms of scientific decision making or judgment. Although the degree and ubiquity of complexity is often overlooked, even judgments that appear to be simple, on closer analysis, are found to contain complexity. For example, certain methodological decisions appear straightforward. But there are innumerable instances in which scientists discovered that methodological procedures they had taken for granted, or considered adequate for establishing experimental control, held unexpected relationships to their research findings and revealed various subtleties and complexities. Pasteur's discoveries provide one example. Any experienced researcher has a wealth of such examples to cite.

No single example reveals the pervasiveness of situations involving complex information and considerations, and recognition of the general presence of complexity is so central to subsequent arguments that a broader examination is required. To do this, one can initially subdivide science using any of a number of available schemes. One might categorize different types of scientific activities, subdivide science into the contexts of discovery and justification, separate logical operations from empirical ones, or order common interpretations or constructions according to their purported level of inference or abstraction. I will take the last of these approaches as a starting point here, with the discussion directed towards two opposing extremes: those judgments supposedly requiring the highest level of inference — theoretical constructions; and those supposedly requiring the lowest level of inference and abstraction — factual or observation statements.

The Complexity of Theories

The presence of complexity is obvious in the case of theoretical constructions. Each postulate of a theory, in part, represents a summary or interpretation of a wealth of data. The data rarely, if ever, provide a simple, consistent, and unidimensional pattern from which one inductively forms a postulate. Rather, the data are usually complex, containing inconsistencies and multidimensional patterns; they must be sifted through, organized, and integrated. And there is often no possibility that postulates can be directly induced from the data, for the most useful postulates usually go beyond or extend the data. Further, theories consist of multiple postulates, which must not only be interrelated with each other, but also with various theorems and observation statements. These relationships can become so complex that individuals have attempted to apply schematic devices to aid in their comprehension, one example being Cronbach and Meehl's (1955) nomological network. Additional components of theories, such as dictionaries and models, add to the complexity of these constructions. Without exception, theories also contain implicit or underlying metaphysical assumptions, such as beliefs about the nature of space and matter. Realizing these things, it becomes difficult to dispute the claim that a complex set of information is relevant to theories.

The Complexity of Factual Perceiving

Although complexity is obvious at the level of theory, it is less obvious for judgments supposedly requiring minimal inference, such as those involved in deriving the most basic observation statements or facts. What is obvious is that facts do not interpret themselves to the scientist, even one who is able to free his or her mind from the mists of biasing thoughts or emotions, but that information must be rendered into intelligible facts through the activities of the inquiring mind. Using Hanson's (1958) phrase to make what is now an oft reiterated point, seeing (facts) is "theory laden." This point is now widely accepted, and the lucid discussions of various scholars (e.g.,

Bunge 1967; Hayek 1952, 1969; Lewis 1929; Sellars 1963; Turner 1967; Weimer 1979) eliminate the need for lengthy coverage here. Rather, quotations from a past and a contemporary writer should suffice. Years ago Goethe issued the famous maxim: "All that is factual is already theory." More recently, Turner (1967) said that

> fact is inseparable from theory. It is an hypothesis, a conjecture
> germinated in theory, which guides the scientist in his search
> for evidence. What kind of world one finds is determined by what
> kind of structure the world is assumed to have. . . . To alter
> a theory, even to reject it, is ultimately to alter, even to reject, the
> fact. . . . Thus the world is no longer flat, nor does the sun
> revolve about the earth, nor do objects in motion seek their natural
> position. Yet each of these at some time was the fact of observant
> men. There are innumerable examples wherein a new conceptual-
> factual context introduces a different excursion into reality. (p. 15)

The awareness that factual seeing is theory laden might be considered revolutionary, but it only serves to indicate that the cognitive activity of the scientist is a problem worth studying. This is only the most preliminary step in uncovering how the scientist proceeds and what the scientist can and cannot do in transforming information into fact. We must move beyond this acknowledgment of factual seeing as theory laden and look more deeply into the problem-solving tasks faced by the scientist—focusing here on the complexity inherent in factual seeing. If complexity can be revealed even at this supposedly basic level, it should become obvious that complexity exists also at higher levels of inference and across scientific activities.

THE SCIENTIST AS FACTUAL PERCEIVER

In various writings about the recognition or identification of facts by scientists, terminology becomes troublesome. The most commonly used term for this activity is *seeing*. The term is used broadly, however, sometimes referring to *cognitive activity* or cognizing, such as when one discusses a world view, and sometimes to *perceptual activity*, as when one discusses the seeing of facts. Unfortunately, it is often not clear what meaning is intended. When referring to the recognition of facts, the use of the term as a synonym for perceptual, as opposed to cognitive, activity appears to provide a closer approximation to the actual mental processes involved. To avoid ambiguity and gain precision, I will substitute *perceiving* for *seeing* in the following discussion. Perception refers to the act of transforming sensory information (from any modality or across modalities) into recognizable objects or "events" (e.g., the sun, or a collision between particles).

What is the task faced by the scientist as factual perceiver? The task of all perceivers, be they scientists or not, is to identify constancies, or stable features, in an environment that presents a nearly infinite number of continually fluctuating stimuli. This is a difficult achievement for various reasons, not the least of which is that the stimuli, taken at face value, are often misleading. For example, change in the stimuli do not necessarily indicate a change in the object. If one looks at a resting coin first at table level and then from above, the stimuli entering the nervous system do literally change, but the coin does not. How are we to distinguish the cases in which changes in stimuli are and are not indicative of a change in the object? We must also distinguish between "signal" and "noise," that is, we must select key elements of stimuli and filter out irrelevant, random, and noninformative details. For example, when we fixate our gaze on an object and then look elsewhere, afterimages of the initial object remain and must be ignored. (We usually do so with such success that we are unaware of afterimages unless we close our eyes following fixation.) As Gibson (1969) describes it, perception is a matter of extracting information from stimulation and of reducing uncertainty created by, or present in, the stimulation that reaches our senses. Perceptual learning is learning to discriminate stimuli that appear similar and to recognize that changes in surface appearance do not always indicate changes in objects—constancy often underlies variance in the superficial stimuli.

The hard-won capacity to go beyond literal interpretation of stimuli, to not be misled by sensation, permits one to identify stable features in the environment and perceive the world in a more veridical manner than would otherwise be possible. This is the seeming paradox of perception—that processing of information, rather than "unbiased" reception, does not distort our perception but permits us to achieve a more accurate view of the external world. Einstein, with his typical perspicaciousness, made a similar point in 1937: "Man seeks to form for himself, in whatever manner is suitable for him, a simplified and lucid image of the world, and so to overcome the world of experience. . . ."[1] As Einstein recognized, this is one reason why the "subjective" element or, more precisely, the perceiver's active re-shaping and reordering of the inflowing information should not be viewed as merely a contaminant. If we could eliminate this form of "contamination" and somehow return to a literal and raw reception of stimuli, we would not perceive the world more accurately and ob-jectively, as some writers seem to think. Rather, we would become stupid and blind—slaves to often misleading superficial appearances. Although our perceptions can be wrong, we will make progress not by eliminating human processing but by improving and extending it.

A central goal of science is to move beyond superficial appearances; more accurately, the goal is to extend the unaided perceiver's capaci-ty to move beyond superficial appearances. This does *not* mean that scientists should return to raw sensory information; rather, as Hayek (1952) has argued, they should attempt to create a "new sensory order." In this new order, direct sensory information is increasingly replaced by information gathered through sense-extending instru-ments, with theory providing guidance about the information that is to be sought and how it is to be obtained and interpreted. As with the unaided observer, the scientist's task is to move beyond super-ficial appearances (in this case, the world's appearance to the unaided perceiver), identifying similarities between objects that appear differ-ent, differences between objects that appear similar, and relationships and patterns that were previously not perceived. In Hayek's view, dif-ferent scientific fields have progressed to various extents in carving out this new sensory order, with the physical sciences thus far being the most successful.[2]

THE COMPLEXITY OF THE SENSORY WORLD

Ordering the sensory world, regardless of the level at which scientists operate, is an arduous task. Scientists within all fields have struggled to derive reliable facts, and these struggles are primarily created by the complexity of the sensory world. As Einstein recognized, the sci-entist is faced with the problem of identifying facts "from the im-mense abundance of the most complex experience . . ."[3] The stimuli are nearly infinite; variations in possible observations and observation-al procedures are infinite. To achieve constancies or reliable scientific facts, one must thus eliminate entire worlds of possibilities, in the process making numerous judgments and decisions that substantially influence outcome. Heisenberg (1972, 62) quotes Einstein as saying that we "must appreciate that observation is a very complicated process."

To appreciate the complexity inherent in factual perceiving, let us first examine the scientist's sensory world and then the decisions needed to order this world. Even if one disregards all of the sensory information that must be eliminated when observing facts, the entities or events from which facts are derived are still complex. These entities or events, whether sensed directly or revealed through instruments, form complex and multidimensional stimuli; they are nothing as simple as the direct and unambiguous sensory experiences described by some past philosophers. We see this most clearly in the social sciences. A psychologist, observing and coding the behavior of a group of five-year-olds, sees one child strike another and puts a check under the column "Aggressive Behavior." Exactly what was relevant to this classification? Was it that the force of the child's blow exceeded a certain level; that the behavior was directed, successfully, toward another child; that there were indications, marked by accompanying verbal statements or specific preparatory movements, that the act was intentional; or that other nonverbal indicators were exhibited? Obviously, "aggressive behavior" is multidimensional, and one must consider not only which dimensions are relevant and how the dimensions are to be measured but also the context in which the behavior occurred. Such complexities may be most evident in the social sciences, given the relatively gross level of basic observations, but they are by no means unique to these fields. Similar complexities are shared by all fields, even physics. Physicists have attempted, without success, to eliminate complexity and ambiguity from perceiving by using highly refined instruments to probe more deeply into the microstructure of phenomena. Hydrogen atoms were inferred from complex patterns of scintillations on zinc sulfide screens. Subatomic structures are complex. Regardless of what one looks at and how carefully one looks, as one looks closer, complex phenomena appear. We may one day find elemental parts and elemental observations, but these things do not exist in contemporary science.

While the events and entities described as facts are themselves complex, substantial additional complexity is created by the variability shown within the class or set of events grouped together as instances of a particular fact. Facts are just this, categories or classes.

They are not single events, but rather sets of events that are grouped together. Although each event that is grouped with other events as an instance of a specific fact assumedly holds one or more features in common with other members of the group, significant variation often still exists among these group members. For example, group members can differ along dimensions considered irrelevant to their factual classification. As Turner (1967) describes it, "The language of fact does not report the *particular* . . . i.e., the particular extension of atomic events, but only that the configuration belongs to a particular class of molecular propositions" (p. 31). Perceiving facts is an act of classification, which introduces a host of considerations: Which features are central, which are secondary, and which are irrelevant to classification? What are the level and types of variation permitted among the events grouped together as instances of a specific fact? How are different facts to be distinguished and classified into different categories and how are their boundaries to be defined? How does one proceed if an observation fits two categories? How are appropriate labels to be selected?

MANAGING THE SENSORY WORLD

The complexity of the sensory world raises many involved issues and considerations, and the conceptual requirements and operations needed to deal with them add to the complexity of factual perceiving. The scientist could not begin to manage the sensory world without a guiding conceptual framework. Theories or, if one prefers, paradigms provide this framework. They serve as guides for determining such things as conditions and measurements (procedures) required for observing relevant phenomena, the features of phenomena to be considered relevant, and the system used for classifying observations. One could go further in arguing, as Weimer (1979) and Spencer Brown (1972) have, that theories include a series of implicit and explicit *commands* for how one should proceed in order to perceive what one *should* perceive.[4] This view implies that the perception of facts requires one to apply the lens of a theoretical system. Perceiving a fact is using a theory, and such perception is not so much theory laden as it is theory dependent. Each perception of a fact may utilize much or all of the conceptual framework of the guiding theory, although this dependence may be obscured when expertise allows almost automatic and effortless factual perception (as described below). "Finished" perceptual achievements often mask the complexity of the underlying processes.

Not only are the theories that guide factual perceiving complex, but learning to apply these theories is also a difficult achievement that requires involved mental operations and decisions. For example, developing competence at application often requires years of education, both to master the theory's frame of reference and to learn to see examples of its application to nature—to recognize instances of the specified theoretical phenomena as they occur in nature. Kuhn (1970) makes this point when discussing exemplars, his term for broadly acknowledged puzzle solutions realized by the application of a theoretical system to nature. Kuhn reports that students are able to conceptualize descriptions of puzzle solutions, but initially they cannot identify phenomena in the laboratory as instances of nature to which these puzzle solutions can be applied. Students are even less able to generalize the specific solutions they have learned to new and unfamiliar phenomena. Only gradually does the successful student acquire these abilities. As Kuhn describes it:

The student discovers . . . a way to see his problem as *like* a problem he has already encountered. Having seen the resemblance, grasped the analogy between two or more distinct problems, he can interrelate symbols and attach them to nature in the ways that have proved effective before. . . . The resultant ability to see a variety of situations as like each other . . . is, I think, the main thing a student acquires by doing exemplary problems, whether with a pencil and paper or in a well-designed laboratory. After he has completed a certain number . . . he views the situations that confront him as a scientist in the same gestalt as other members of his specialists' group. (p. 189)

Finally, in addition to the mental operations that must be performed, one can briefly examine some of the issues and considerations that must be dealt with by the scientist as factual perceiver. The area of measurement can be used as a representative example. Measurement procedures raise broad and numerous questions. For example, what temporal span should each act of measurement occupy? How is the unit of measurement defined? Should we use our senses, sense-extending instruments, or other forms of measurement?

We know we alter what we measure, but what forms of alteration can be expected, which can be avoided, and how can we avoid them? How do we proceed when an entity is in continual flux? These are anything but trivial questions that can be casually dismissed, although their relevance and impact are usually less visible when conventional procedures have been broadly accepted.

Additional complexities could be introduced, but the point is already clear. Perceiving facts requires the consideration of complex information and involved issues. Facts describe entities and events that are complex and multidimensional. The context within which these entities exist or these events occur must be considered. Observations must be classified into factual groups and decisions must be made about measurement, both of which require higher-level judgments. Theories, which provide guidelines for making these decisions, include not only a set of implicit and explicit assumptions but also a broad set of extratheoretical assumptions. The full armamentarium, then, is used in factual perceiving. Constructing and evaluating facts often recapitulate nearly the full range of acts necessary for constructing and evaluating theories, two activities that are closely interrelated and generally occur in unison. For this reason, describing factual perceiving as lower-level inference is misleading because it ignores its parallels to and dependence on the higher-level inferences involved in theoretical judgments.

It is apparent, then, that even scientific judgments supposedly requiring minimal inference necessitate the consideration of complex information and complex issues. This complexity may be obscured, especially when extended practice and expertise create the capacity for seemingly effortless factual perception, or when consensus in the community permits scientists temporarily to ignore the many subtleties and underlying issues. Underlying complexity immediately becomes apparent, however, when new phenomena are uncovered that do not fit into the framework of the world view, or when theories change and "facts" change with them. In the former instance, a variety of involved conceptual activities suddenly come to the forefront as attempts are made to explain a new phenomenon—to fit it within a category of the current conceptual framework or to determine what fact it represents. An excellent example is provided by Zenzen and Restivo's (1982) report on a group of colloid chemists. Colloid chemistry involves the study of mixed or intermediate states between liquids, solids, and gases. One of the chemists discovered an immiscible liquid, a finding incongruent with certain assumptions normally taken for granted by this group. The observation of this previously unobserved dispersion between liquids evoked a series of investigations and a plethora of theoretical and methodological arguments aimed at clarifying and categorizing the phenomenon. It was only when the available system could not be applied to a new observation in the effortless and automatic way it was applied to familiar observations that the complexity of the underlying framework and considerations was revealed.

The complexity of factual perceiving is also revealed when world views or theories change. When theories change, facts change also. As Harré (1972) states:

Descriptions of observations cannot be entirely independent of theory either in form or in content. There are no modes of description which remain invariant under all changes of theory. The way in which observations are described changes when theory changes. (p. 25)

The restructuring of facts requires scientists to learn a new theoretical system and to relearn how to interpret or perceive phenomena. The complexity of this process makes evident the complex information and considerations that enter into factual perceiving. For example, shifts occur in beliefs about the observations that are to be considered credible and those that are to be dismissed as artifacts. Beliefs about methodology and measurement often change, as well as the labels attached to classes of observations. Again, Kuhn (1970) provides a useful example. Before the chemical atomic theory of Dalton, obtained weights of composite elements in compounds were often nonproportional. Dalton's theory required fixed proportions for any *true* chemical reaction. The change in theory created a substantial alteration in perceiving. In Kuhn's words, "Here and there the very numerical data of chemistry began to shift. . . . When it was done, even the percentage composition of well-known compounds was different. The data themselves had changed" (p. 134-35).

Overview and Extension of the Argument:
The Pervasiveness of Complexity

I have made an argument for the presence of complexity not only at the level of theory but also at the level of factual perceiving, or of scientific judgments supposedly requiring minimal inference. Perhaps it should be enough to say that complexity is present across scientific activities, given that most or all of these activities require judgments at a level of inference that equals or surpasses those required for factual perceiving. However, the adequacy of such an abstract generalization is questionable. Historical, philosophical, sociological, and psychological literature is replete with failures to recognize the complexity of scientific activities and (over)simplified descriptions of science. Therefore, it seems necessary to survey basic types of activities or facets of science and to point out their inherent complexity.

Any strict division of science into its various facets is artificial because the categories are continuous rather than discrete, but division can serve as a conceptual aid so long as one recognizes its limitations. I will use the following strategy of division: at the broadest level, science is subdivided into method and knowledge; for present purposes, method is itself divided into inquiry (activities involving the collection of information) and assessment (activities involving the evaluation of information). No further subdivision of the scientific knowledge base will be made at this point.

INFORMATION GATHERING

How complex is inquiry or information gathering? Beginning with measurement, one can look at the complexity of possibly the most basic operational act—counting occurrences. Even if counts are conducted by simply recording the numbers an instrument provides, the interpretation of results depends on the application of higher-level constructions, which address such matters as the conceptualization of the thing being measured and the validity of the measurement procedure. Because measurement cannot be separated from or exist without factual perceiving, it encompasses all the complexities of factual perceiving that have been discussed above. Even the most basic measurement, however, requires more than factual interpretations. It is also dependent upon a broad range of assumptions about the performed operations. Some of these assumptions address questions pertaining to the more technical aspects of procedure: What is the best way to operate measurement instruments? What environmental conditions must be held constant? What variables or forces must be prevented from impinging upon the thing being measured? Other assumptions address basic metaphysical questions: Are entities continuous or discontinuous? Are they reducible to more essential elements? To what extent can an entity change and still be assumed to maintain the same identity?

If basic measurements involve numerous complexities, then higher-level measurements obviously do also. For example, one often attempts to measure variables that have no direct referents and can only be inferred. Sometimes there may be no grounds for assuming that the variable being "measured" actually exists, and there may be no need to make this assumption. Hypothetical constructs often cannot be demonstrated directly, nor do they necessarily require ontological claims. Nevertheless, they can still serve a useful function by going beyond the data and suggesting how and what one should measure.

A physicist, Martin Deutsch (1959), provides a description of the laboratory environment that poignantly illustrates the complexities involved in measurement:

A typical high-energy physics laboratory abounds in impressive sights and sounds. . . .

I am sure that a scientist of 150 years ago, told to proceed with experimentation in this laboratory, would have no difficulty in making extensive observations on the change in the appearance of the lights as various knobs are turned, and on the variation in the pitch of the generating noise. But, of course, none of these changes directly accessible to sensory impression are really relevant to the experiment actually being carried out. . . . The only visible feature intimately connected with the actual experiment may be the position of a knob controlling the current through a magnet and a mechanical recording device indicating the rate of arrival of electrical signals from a small counter. . . . Both the change made—the position of the control knob—and the resulting effect—the reading on the recorder—seem almost negligible in the totality of material involved in the experiment. . . .

. . . To make matters worse, there are probably several other knobs which could be turned with much more drastic effect on our recorder. In addition, there are literally dozens of similar recording devices and hundreds of other similar knobs connected with the experiment, some of which may show correlations much more marked than those under investigation. . . . We see, then, a situation in which . . . a large number of related phenomena remain uninvestigated. . . .

. . . in a typical experiment in modern physics the apparatus

involves complex influences frequently of much greater order
of magnitude than the phenomenon investigated. Each of these influences can occupy a lifetime of experimental investigation before
it is fully understood. . . . (pp. 98-102)

Measurement is only one facet of information gathering. A related
facet, experimental design, itself contains numerous aspects, and my
discussion will be directed towards only one of these—the selection
of research strategies. Researchers must select from many generic
strategies, ranging from natural observation to rigorously controlled
experimentation. Numerous variations of each generic strategy are
also available. Even Campbell and Stanley's (1963) brief introductory
book describes more than twenty different research strategies. There
are no ultimate rules for the selection of strategies, as witnessed by
the hundreds of frequently conflicting articles written on this topic,
and convention and pragmatic considerations dictate choice. Selection of strategy, therefore, requires judgment. Relevant considerations, at minimum, include one's beliefs about the phenomenon of
interest, research goals, underlying assumptions about the strategies,
and evaluation of the match between the phenomenon of interest
and strategy. Further, one must know both the strategies and how to
apply them. Mastery is probably obtained in a fashion similar to the
two-step process involved in learning exemplars and then learning to
apply them to nature. One learns research strategies, and then,
through practice, learns when and how to apply them to nature. One
learns to see essential similarities between familiar phenomena that
have been studied successfully using particular research strategies and
new phenomena to which these strategies have not been previously
applied. In this way, one learns to extend the strategy to new puzzles
or questions. The need to learn both the strategies and their application to nature increases the complexity of the judgments required for
this aspect of information gathering.

Judgments pertaining to research strategy go beyond issues of selection and application. Strategies form part of the context within
which results are interpreted. They influence judgments about the
confidence one can place in particular results, the factors that are
considered possible confounds, the alternative explanations that may
account for results, and the type of relationships that are uncovered
between variables. For example, many psychologists are quick to
point out that "uncontrolled" clinical observation is a preliminary

step in the acquisition of knowledge, that these observations are biased by the observer's model of the mind, and that the "data" are contaminated by the manner in which the clinician interacts with the patient. Further, the same psychologists who readily jump to causal explanations on the basis of clinical observation may exercise considerable caution before assuming causal links when using another strategy to study the same phenomenon.

As one can infer from the above, the selection of strategies may not only reflect but may also shape one's world view by providing a context for interpretation, particular forms of information, and even a model of nature. For example, Bowers (1973) argues that the world view of behavioral psychologists has been colored by the specific research strategies they use. According to Bowers, these strategies, which involve the strict study of antecedent conditions and subsequent behavior, have become a model for human behavior, leading to the view that human responses are causally linked to antecedent conditions. Cronbach (1957) makes a similar argument when discussing a group of psychologists who placed heavy reliance on correlational measures:

The correlational psychologist was led into temptation by his own success. . . . A naive operationalism enthroned theory of test performance in the place of theory of mental processes. And premature enthusiasm exalted a few measurements chosen almost by accident from the tester's stock as the ruling forces of the mental universe. (p. 675)

In the examples discussed by Bowers and Cronbach, the precise relationship between strategy and world view is open to debate, but it is certainly possible that in these and other cases the relationship is bi-directional.[5]

One additional complicating factor can be mentioned. Although it is now broadly accepted that studying an object always changes that object to some degree, different strategies of investigation may exert different effects upon objects, something that should be considered when evaluating the results obtained through the use of specific strategies. For example, the gender of an experimenter may have little impact if an anonymous questionnaire is used to study deviant sexual behavior but substantial impact in face-to-face interviews. These effects and problems may be most obvious in the social sciences, but they are also of major relevance in the hard sciences, as often becomes clear when unexpected observations or anomalies are uncovered.

ASSESSMENT AND THE KNOWLEDGE BASE

The complexities inherent in measurement and research strategy are representative of the complexities involved in information gathering. These examples were not selected because of their unusual complexity; any other facet of information gathering could be shown to be similarly complex. The same is true for the second subdivision of methodology, that of assessment, and for the scientific knowledge base. In the area of assessment, the scientist may evaluate anything from facts to theories, and few (if any) judgment situations are simple. When evaluating or assessing facts, as we have seen, complex information and considerations are relevant. For example, facts describe classes of complex and varying events or entities. Assumptions about measurement and the nature of the things being measured should be taken into account when designing assessment methods. One must also apply a complex theoretical framework. All of this information and these considerations, plus additional information and considerations, are relevant to the assessment of hypotheses, theorems, and postulates, making such judgments even more complex than those required for factual assessment. No simple programs for the evaluation of facts, hypotheses, theorems, or theories should mislead one into assuming that this process is simple and direct.

Popper's (1959) critical rationalism provides an example of a program for evaluating theories that, on the surface, appears simple and straightforward but, in practice, is not. Feyerabend (1979) makes this point forcefully, using as a literary device an imaginary conversation between the supporter of critical rationalism and a protagonist. The latter states:

Popper . . . says it is rational to reject a theory that conflicts with an *accepted* fact, it is irrational to retain it. And he assumes that such a directive *suffices* to make theory selection unanimous and debates about it rational. Now it turns out that whenever somebody rejects a theory on the basis of accepted refuting evidence he makes assumptions which are just as rational (or irrational) as the assumptions of his opponent who refuses to accept that evidence as refuting. Theory choice is still not rational *because the moves that would make it rational rest on assumptions which are often nothing*

but gut feelings: First the parties have to decide what evidence they
will accept as refuting evidence, *then* the 'rational procedure' of
refutation can start. But the decision involves elements that are no
longer rational because they concern the circumstances under which
this particular form of rationality becomes effective. The situation
is even worse. Assume all the parties have accepted some facts as real
and the facts contradict a theory all of them share. Then it is still
possible to postulate unknown effects, 'hidden variables' that are
responsible for the appearance of a conflict when there is actually
no conflict. Again such an assumption is as 'reasonable' as the
assumption of the absence of all such effects and again the latter
assumption is presupposed when we say that the theory has finally
been rejected in a 'rational' manner. We can of course imagine
a world in which Popper's methodology gives a thoroughly rational
account of *all* our moves. Such a world would consist of situations
that can be identified in a relatively simple manner, there are no
disturbances of our experiments and our observations, facts are never
mixed up with each other, there are no 'hidden facts'. . . . I
think it is clear that we don't live in such a world. . . . unfortunately, people seem to prefer simple rules to a view that emphasizes
the complexity of all decisions. (pp. 98-99)

The knowledge base of a field is the last facet of science to be discussed. The knowledge base is defined somewhat narrowly here as
the data and broader constructions of data—be they hypotheses,
theorems, postulates, or theories—that are inherent to a scientific
field. It is hardly necessary to reiterate that the body of knowledge
contains complex information. The complexity of theories and facts
has already been discussed, and the same arguments concerning such
things as theory-based perceiving, the need to select and integrate
multidimensional information, and problems in classification also apply to other aspects of the knowledge base.

In summary, regardless of the supposed level of inference and the
type of activity or facet of science with which they are involved, scientists are faced with a complex of relevant data and considerations.
The closer one looks at these activities and facets, the greater are the
complexities that one discovers—even when science is divided into
specific facets. Virtually every individual act or judgment within each
facet is complex, and actual scientific practice requires the integration of these facets. For this reason, science as a whole is a far more

complex enterprise than consideration of the individual parts would indicate. These parts are complex in themselves, but when taken together the whole is of substantially greater complexity.

The Hidden Nature of Complexity

When one views science as a whole, its complexity is apparent; when one views the components of science separately, however, their complexity is often less obvious. Of the many possible reasons why forms of complexity have been overlooked, I will offer brief speculation about three bases.

First, one suspects that earlier views of science, which depicted the enterprise as far simpler that we now see it, have not been completely abandoned or unlearned. The view of science as an intellectual, constructive (but not irrational) activity, a view towards which we have been moving, is more conducive to the discovery of complexity. As it becomes increasingly clear that knowledge is not a simple compilation of simple facts, it becomes increasingly clear that knowledge acquisition requires complex judgments and decisions. One can question, however, whether this new view has been fully extended.

While there is broad recognition of the complexity inherent in some scientific activities, especially those considered higher-level inferential tasks, it is often not recognized that complexity characterizes virtually all scientific activities, even those considered lower-level inferential tasks. Stated in another way, we have been gradually adopting a new framework or paradigm for viewing science. As is generally the case, learning to apply a new view, or to recognize phenomena as instances of the new view, is a developmental process. Over time, the view is increasingly extended. Our new view of science has perhaps not been fully extended; upon broader extension, the substantial complexity inherent across scientific activities and judgments may become more apparent.

Second, we also may be misled by our introspections. There is reason to suspect that many of the cognitive operations scientists perform in interpreting the world take place outside of conscious awareness. One need not adopt a psychoanalytic model of the unconscious mind to make this argument. For example, as Newell and Simon (1972) have done, one can postulate that through repeated practice cognitive operations that are initially learned and applied consciously can be executive automatically and with little, if any, need for conscious mediation. For example, during the earlier stages of their education, future chess masters must often analyze board patterns methodically. Eventually, with repeated exposures to a variety of patterns, many are rapidly recognized and require minimal conscious analysis. (This may represent the transformation of an essentially cognitive process into a perceptual one.)[6] Scientists may undergo a similar transition as they acquire expertise. Early attempts to apply world views are methodical and plodding, but over time they become effortless and automatic. Application may become so automatic that the underlying chains of reasoning are inaccessible to conscious analysis. The common inability of scientists to verbalize the reasoning steps underlying their interpretations may partially reflect such a process. Perhaps what Polanyi (1966) calls tacit knowledge, or what Kuhn (1970) considers rules of seeing that cannot be verbalized, actually reflect cognitive operations that were once conscious but that have become automatic and inaccessible to conscious analysis. Scientists, then, may often be unaware of the cognitive strategies they use, precluding reliable phenomenological reports. Although scientists sometimes may overestimate their capacity to perform certain types of complex cognitive operations (see chap. 4), at other times they may fail to recognize many of the reasoning and perceptual processes that underlie their interpretations of the world.

Third, scientists may often fail to consider more than a relatively limited range of information. Attention to the information that they do consider and the relatively simple conclusions that they draw, rather than to the information that is available, may lead one to erroneously assume that complex information is not present. In other words, the range of information that is actually considered and managed should not be mistaken with the range of relevant information that is potentially available and could be considered. In examining this possibility, we should recognize what has and has not been claimed thus far. It has been claimed that for most or all scientific judgments, a complex matrix of relevant information and considerations exists. It has not been claimed that, in practice, scientists take all of this information or these considerations into account, or at least that they do this competently. (In fact, the central thesis of this work is that they are incapable of doing so.) In practice, scientists may often use cognitive and problem-solving approaches that simplify decision making and potentially mask the presence of complex information. Their use of these strategies does not indicate that the available information is simple; rather, it may represent a response to the presence of complex information.

Scientists appear commonly to use reasoning strategies that simplify decision making. The mechanisms and factors underlying such strategies will be discussed in chapter 4; for now, I will point to possible examples of simplifying strategies. Themata, as described by Holton (1973), may be one example. According to Holton, themata are a limited set of fundamental ideas, or guiding conceptual frameworks, which are used by scientists in numerous fields. Most themata serve as organizing constructs that aid in translating complex phenomena into a familiar and relatively simple world view. Concepts such as symmetry, continuum, and hierarchy are instances. By simplifying complex phenomena, themata allow one to comprehend what might otherwise be incomprehensible. As Holton (1978) puts it, "Helping to make sense of the world in a way not possible through the demands of logicality alone is indeed one of the chief functions of a thema" (p. 16). Holton believes that the use of themata is extremely common among scientists.

Holton (1978) illustrates the presence and function of themata by drawing upon the attempts of physicists to order the "bewildering" variety of elementary particles. The theme of hierarchy serves this ordering function: "There are ways of . . . mastering the bewildering variety. The ordering of chaos by means of the concept of hierarchy or levels of categories—a manageable few, just four—comes to the rescue as a methodological theme" (p. 16). The levels of the hierarchy represent types of interactions among particles, or forces (e.g., electromagnetic interactions). As Holton mentions, the use of the hierarchy is obviously far from original or unique: "The theory evokes more than an echo of an older scheme of fourfold categories, one so magnificently successful that it helped to rationalize the observable phenomena for some 2,000 years: the four Elements, with their own internal hierarchy . . ." (p. 16). But the goal of the physicists is to identify previously unrecognized unity among particle forces: "The new unification through hierarchical ordering promises . . . that two and perhaps three of the forces in the four categories 'have an underlying identity'" (p. 16).[7]

A second example of a simplifying strategy may be the tendency to study isolated phenomena or to restrict investigation to a very limited domain. This tendency is reflected in both the high level of specialization found within scientific fields and the popularity of analytic approaches. Both serve to reduce the range of phenomena that require consideration, thereby simplifying the scientist's task. Newton said that "the Investigation of difficult Things by the Method of Analysis, Ought ever to precede the Method of Composition" ([1704] 1931, 404). Newton apparently believed that such a strategy was necessitated by human cognitive limitations. As beginning students commonly show by proposing to study impossibly broad problems, learning to do science is, in significant part, learning to restrict the domain of inquiry.

As a third example, many prescriptive programs for science can also be viewed as simplifying strategies. Prescriptive programs, at least in theory, aid the scientist by providing problem-solving strategies that simplify decision making. These programs often are modified or eventually rejected, not because they are illogical but because they are found to be oversimplifications. Their application, which is often at the level of thought experiments, invariably leads to the discovery of important problems they do not help solve; this recognition of complexities that these strategies cannot manage commonly leads to their modification or rejection. For example, disconfirmation is now seen as insufficient by itself, not because it is illogical but because it does not address many of the decisions scientists must make.

Although the presence of complexity can be easily overlooked, its signs and manifestations are revealed in many common scientific occurrences. The variability or unreliability of judgment may be one manifestation. Kuhn (1970) offers relatively strong documentation for his claim that scientific schools have extraordinary difficulty reaching consensus. This difficulty can be seen, in part, as a reflection of the complexity of information, which makes its organization and interpretation extremely difficult and leads to a wide variety of opinions or positions. Were the information simpler, such diversity in thought would be unlikely, a possibility that will be further examined in chapter 4. The processes that take place when a crisis or change in interpretation occurs clearly reveal underlying complexity. When the system of givens can no longer be uncritically accepted, attention is directed to a broad range of considerations. These occurrences reveal the complex network of assumptions that otherwise may remain largely underground, as scientists debate anything from broad metaphysical assumptions to the finest technicalities of measurement. This process can occur on a large scale, such as in the paradigm shifts that Kuhn has described, or on a relatively small scale, such as in the reactions to single anomalous discoveries that Zenzen and Restivo (1982) have described. Complexity is also revealed in the broad changes that occur once viewpoints are altered. Changes in measurement, research methods, facts, theories, and even basic world views reflect the complexity of the information base and assumptions underlying all scientific constructions.

COGNITION AS A NECESSARY LINK OR FINAL COMMON PATH

Problems of Acognitive Views of Science

The assumption that human cognition plays a central role in science seems to follow naturally from the assumption that science is a constructive endeavor requiring complex judgments. Many individuals, however, accept the latter but not the former assumption, leading to descriptions of science that are "acognitive" or that place cognition in a secondary role. In acognitive descriptions, the role of constructive human thinking is either ignored, treated as an epiphenomenon or variable that contributes nothing unique to scientific knowledge, or viewed as mere noise or distortion. The acognitive view can be traced back at least to Bacon, who believed that the facts could speak for themselves and that any contribution of the knower to the known represented distorting bias. The acognitive tradition was carried forth by early empiricists and by the logical positivists, who believed that the scientist had the potential to be a completely objective recorder and compiler of sensory data. Ideally, cognition mirrored external reality, and any constructive contribution was seen as a form of undesirable distortion. It was as if the "data" could produce the facts and did not have to be processed through the human cognitive machine.

Virtually all modern thinkers have abandoned idealized models of the scientist as direct receiver of external reality, accepting instead the view that scientific thinking is shaped by more than data alone. Some sociologists, however, have taken up the acognitive tradition, albeit in a different form. Two variants of this modified acognitive position appear and will be discussed in turn. The first, which is now waning in popularity, is the extreme view that scientific ideas are a direct product of social factors or social conditions, such as conditions of labor. (See Böhme 1977 and MacLeod 1977 for overviews of this and the second variant discussed below.) This position can be labeled social macroreductionism, or, if one feels less conciliatory, sociocentrism. The roots of modern social macroreductionism can be found in the works of earlier sociologists of knowledge, such as Durkheim (1915), who took the position that knowledge was relative to, or determined by, social conditions. It was but a small step to apply the same argument to science. Scientific belief became a product of social factors and perhaps data, and the importance of human cognition was minimized or totally dismissed.

The social macroreduction of science can be challenged on many points. When taken at face value, it leads to an obvious dilemma. These sociologists acknowledge that scientific belief is not a direct product of data; rather, the data pass through an interpretive sieve. They argue that the manner in which this sieve shapes beliefs, or the manner in which the data are interpreted, is determined by social conditions or factors. However, social factors can themselves be conceptualized as forms of information, a point that will later be developed in detail. One cannot argue that scientific information *does not* determine belief directly, but must be interpreted, while at the same time implicitly or explicitly arguing that social information or conditions shape belief directly and themselves are not subject to interpretation. To my knowledge, no one has provided a satisfactory explanation for this apparent contradiction. Social information does not enter the mind through osmosis, and belief is not a direct mirror of social information. Therefore, one must be able to account for the factors that shape the impact of social factors on belief. The account provided by social macroreductionism is hardly sufficient. One has recourse only to social factors as shaping influences and, therefore, must argue that the impact of social factors is determined by higher-order social factors. Taken to its logical extension, the impact of higher-order social factors must be accounted for by even higher-order social factors, which must be accounted for by still higher-order social factors, and so on. One is led towards an unending progression that provides no explanation for the variables that shape the influence of social factors.

Further, if one takes social macroreductionism literally, then the role of *individual* factors — individual human cognition included — must be considered irrelevant because it is assumed that information and social factors alone shape scientific knowledge. This creates major problems: for example, one would be forced to take the position that a mental retardate, exposed to the identical social conditions and "data" as was Einstein, would generate the theory of relativity. If one counters that the retardate could not possibly be exposed to the identical social circumstances because he or she would be incapable of taking part in many of the social interactions that were formative to Einstein's thoughts, then one has admitted to the role of individual factors and has, thereby, abandoned strict social macroreductionism. Moreover, given the assumptions of strict social macroreductionism, all individuals exposed to similar information and social conditions would develop similar beliefs. Any difference in belief, therefore, must be reduced to differences in personal history. This is similar to a position maintained by many behavioral psychologists who argued that individual differences in behavior can be accounted for by differences in past environmental conditions or reinforcement history. This position has caused the behaviorists major embarrassments. For example, it has been firmly established that even when monozygotic twins are raised apart, if one twin develops schizophrenia the other is very likely to develop the disease also (e.g., Gottesman & Shields 1972). Individual factors, rather than environmental ones, clearly seem to account for this finding. Explanations based solely on environmental or social factors have been soundly discredited in psychology, and there is no reason to believe that such explanations will experience a different fate in the sociology of science.

These and other arguments have created damaging problems for social macroreductionism (for additional examples see Hayek 1955; Popper 1950), and it is not surprising that its popularity has steadily declined. In its wake have come a series of positions that make more moderate claims for the importance and role of social factors but that still contain many residuals of the acognitive view. For example, after reading an earlier draft of this manuscript, one reviewer categorically stated that human cognition is of secondary importance when compared with social factors or processes.

I will discuss the relation between human cognition and social factors later in this and subsequent chapters. For now, however, I would point out that arguments for the primacy of either social or cognitive variables are reminiscent of two children quarreling about whether the heart or brain is more essential to life. Obviously, both are essential and both share this common property, but both do not perform identical functions. Is human cognition, or are social processes and factors, most essential to science? Both are essential to scientific life and share this property, but both do not perform the same functions or affect science in the same manner. Therefore, arguments that social variables or that cognitive variables play an essential role in science do not necessarily represent competing claims. Acknowledging or demonstrating the importance of social variables provides no grounds for minimizing or dismissing the importance of human cognition, and vice versa. Global claims to primacy are unproductive, and it would be more useful to study the role and unique contributions of each. Human cognition is unique in being the only mechanism by which information, be it "internal" or "external" to science, is rendered meaningful by (to) human beings. Human cognition, then, is *the* medium through which scientific knowledge is constructed.

Conceptualizing Interrelationships among Cognition, Social Factors, and Data

Arguments for the role played by human cognition can be concretized by offering a framework for conceptualizing the relationship among cognition, social factors, and scientific data. One approach, as previously implied by the description of social inputs as information, is to conceptualize both data and social inputs as factors that either provide information or have an impact upon information. It is obvious that scientific data provide a form of information; however, social factors can be seen in a similar fashion. Some social factors provide direct or indirect forms of information, and others influence access or exposure to information. Social inputs that provide direct forms of information include ideas or ideological systems that are external to (arise outside of) a scientific school. For example, dominant sociocultural ideologies are forms of ideas and can be considered direct information. Other sociological forms of ideas or direct information might include contemporary philosophical schools of thought or the extrascientific ideologies of co-workers. These sources of information may have a primary impact on scientific belief, such as when a sociocultural ideology becomes a model for interpreting scientific results, or a secondary impact on belief, such as when a dominant philosophical system influences the choice of research problems.

Indirect forms of information generally stem from responses of others that do not directly communicate ideas but that represent actualized or potential consequences resulting from the actions or ideas of scientists. Grants; scientific awards; social rejection; the admiration of colleagues; and institutional policies dictating hiring, firing, and promotions are all examples. If one wishes, these can be considered social rewards and punishments. As such neobehaviorists as Bandura (1974) have argued, the impact of social rewards and punishments is determined by their information value. They provide information about the actions or ideas that will be rewarded or penalized, information that often serves as a basis for directing or modifying one's behavior. An extreme example is Galileo's public renouncement of his scientific beliefs when threatened with torture or death.

Actual or anticipated social contingencies can influence the beliefs of scientists in numerous ways and through numerous mechanisms. For example, they may affect exposure to information. Perceived consequences may lead a scientist to follow his or her corporation's dictate to study a particular problem. This research might result in discoveries that provide new and direct information that modifies the scientist's beliefs. Social contingencies may also determine access to information, such as when a scientist's efforts are rewarded by funding for new instrumentation that provides increased capacity to measure faint phenomena. At other times, social contingencies and subsequent behavioral changes influence belief in more subtle ways. For example, there is evidence (e.g., Kelley 1967; Bem 1972) that even when one modifies behavior in response to externally imposed demands, such as institutional policies, the very act of engaging in new behavior modifies belief. Assumedly, one's own behavior becomes a source of information that modifies self-attributions and, thereby, one's beliefs (see Bem 1972, for a more detailed explanation). These examples certainly do not exhaust the possible means and mechanisms by which contingencies influence belief. Unfortunately, our knowledge in this area is limited, and further investigation would likely yield important insights into the shaping of scientific belief.

Both data and social factors, then, can be conceptualized as variables that either provide information or have an impact upon exposure or access to information.[8] Although information is the medium through which both exert their affects, it is not the final path through which knowledge is constructed. All forms of information must be interpreted or constructed. It is the interpretation or construction of information that ultimately determines its impact (although construction *is* partially determined by the information). An exaggerated example will illustrate this point. As mentioned, certain social factors provide indirect forms of information about potential or actual environmental contingencies, as the following case illustrates. The head of a laboratory recognizes that she has been remiss in interacting with and encouraging her junior scientists, and she decides to offer words of support to each of them. Jones, who has been considering a job change because he feels unsupported, is reassured by his supervisor's words and decides to remain with the laboratory. Smith, an individual with paranoid trends, has had no thoughts of leaving. But upon hearing much the same words as Jones, Smith immediately interprets this change in his supervisor's behavior as an indication that he is being set up to be dismissed. Rather than face this imminent humiliation, he resigns and cannot be talked out of this decision. Obviously, the difference in the two men's behavior was not a product of the differences in the information to which they were exposed but in their interpretatation of the information.

Although individual variation provides obvious examples of the role interpretation plays in determining the impact of information, one should not assume that the role of interpretation is of lesser importance when consensus is exhibited. In general, when the judgments of individuals are in agreement, there is a tendency to believe that the information itself has determined these judgments and that interpretation has played an insignificant role (Kelley 1973). In a like manner, when scientists evidence agreement, there may be a tendency to assume that the data or information has determined their conclusions and that interpretation has played a minimal role. But this cannot be assumed. Interpretations or constructions can be shared by group members, but this does not make the role of interpretation any less important—only possibly less obvious. For example, the importance of interpretation becomes clear, even among scientists who have achieved consensus, when new findings lead to a change in viewpoint. When the new viewpoint emerges, the old data are seen differently. Such a shift occurs because interpretation plays a central role in the construction of data, regardless of whether consensus does or does not exist. As philosophers of science argue, one cannot hope for pure objectivity because data or information must always be constructed.

Data and social factors, then, can be viewed as the major sources of information contributing to scientific knowledge. Social factors can also influence access and exposure to information, but this information never determines knowledge directly. No matter what its source, information must be interpreted or constructed. The construction of information is a cognitive act. Therefore, human cognition is the final path in the construction of scientific knowledge: It is the factor that ultimately mediates the impact of information.

THE CENTRAL TASK OF THE PSYCHOLOGY OF SCIENCE

If cognition is essential to all scientific knowledge, providing the unique means for its construction, then the study of cognition is essential to the study of science. Psychology is the discipline best suited to the study of cognition, and the central task of the psychology of science is to provide insight into human cognitive processes and the cognitive construction of knowledge. This argument certainly does not underestimate the importance of historical, philosophical, and sociological study of science. These disciplines deserve most of the credit for the advances we have made in understanding science, and the contributions of psychology have thus far been limited. But the study of science seems to require the efforts of many disciplines, with psychology's greatest potential contributions resting on its unique suitability to the study of cognition.

Seen from a broader framework, science is one of a number of epistemological systems for generating knowledge. The study of epistemology has been increasingly pursued by scientists, especially cognitive psychologists (e.g., Piaget 1970, 1971). The psychological study of epistemology can make two general contributions. First, it can serve as a means for cross-validating certain of the descriptive claims that appear in epistemological models addressing the generation of scientific knowledge. Historians, sociologists, and philosophers who try to describe the acquisition of scientific knowledge invariably make implicit or explicit assumptions about the psychological mechanisms underlying this process. For example, Kuhn (1970) assumes that perceptual transformations underlie shifts in viewpoint. Such descriptions of knowledge acquisition, then, contain psychological claims, which often can be tested through the psychological study of human cognition. Weimer (1974a) strongly emphasizes this point when discussing rival attempts to describe the methodology of science.

No methodology that is incompatible with the psychological sciences' account of a scientist's cognitive functioning can be considered correct. Congruence with explanatorily adequate cognitive psychology . . . is a . . . metacriterion with which to assess the rival methodologies of scientific research. (p. 376)

Although Weimer's statement may be somewhat extreme, it does point out the critical contribution that the psychology of science can make in testing or evaluating descriptions of scientific knowledge production.

Descriptions of how scientific knowledge is produced must meet one essential criterion of plausibility—they must reflect the actual cognitive operations used by scientists. The most up-to-date knowledge of human cognition should be used to determine where descriptions stand along this criterion. For example, if cognitive study demonstrates that the factual perceptions of scientists shift when their basic assumptions change, then a description maintaining the occurrence of such shifts should be considered more plausible than one that denies them, at least relative to this claim. It follows from this general criterion of plausibility that descriptions of the cognitive operations or performances of scientists must not exceed the bounds of human capability. Many accounts of science seem to describe cognitive operations that scientists cannot possibly perform, given what we know about human judgment, and these accounts may not pass the test of plausibilility (see chap. 4). The requirement that descriptions stay within plausible human limits sets a fundamental boundary condition on descriptions of science.

In discussing boundary conditions in later chapters, primary attention will be accorded to individual judgment. This may offend those sociologists who claim that the group can do what the individual cannot and that individual limitations thus do not set any boundary conditions. Blind retreat to the assumed emergent abilities of the group seems indefensible, however, because at least some of the assumptions about the emergent capacities of the group can be translated into empirical propositions and subjected to scientific test. Simply to argue that the group has emergent capacities and to dismiss the need for empirical tests represents little more than an appeal to dogma. Moreover, there are strong grounds to argue that individual limitations do set relevant boundary conditions. The potential emergent capacities of the group are still determined, in part, by the properties of the individual parts—by the properties of individual human cognition. New properties may emerge when parts are combined, but the properties that *can* emerge are still limited by the properties of the individual parts. For example, although organic material may be combined to form a living organism with properties that the individual parts do not hold, one cannot make a tree out of granite stones. Similarly, the cognitive capacities of a group of scientists are limited by the cognitive capacities of individual scientists. There are no grounds, then, to assume that the boundaries set by individual cognitive limitations are irrelevant because the group may exhibit certain capacities that individuals do not.

In addition to its role in cross-validating descriptive claims, the psychological study of epistemology can make a second important contribution. It can provide information that is useful in generating and testing prescriptive programs for science. All prescriptive programs make either direct or implicit assumptions about human cognition. For example, they assume certain cognitive capacities, including anything from the ability to apply rules of logic to the ability to integrate multidimensional information. Many of these assumptions can be reworked into empirical propositions that are amenable to scientific testing. Piaget (1970) makes this point in a statement that applies to both description and prescription:

All epistemologists refer to psychological factors in their analyses, but for the most part their references to psychology are speculative and are not based on psychological research. I am convinced that all epistemology brings up factual problems as well as formal ones, and once factual problems are encountered, psychological findings become relevant and should be taken into account. (p. 8)

Some individuals, the majority of whom are philosophers, argue that scientific findings are irrelevant to philosophical prescriptions. There is a tendency to claim special status for philosophical knowledge, to place it above any other form of knowledge or above accountability to scientific findings. At times, there appears to be clear justification for granting this status to philosophical prescriptions for science. Sociologists and psychologists of science have not infrequently suggested broad prescriptive revisions on the basis of minimal evidence or questionable theories. Kubie's (1954) attempts to apply psychoanalytic constructs to the mental processes underlying the activities of scientists and to draw prescriptive implications from this application is one such example. Even Mahoney, who has demonstrated sophistication in his work on the psychology of science (1976), had to be reprimanded by Weimer (1977) for drawing general prescriptive implications from a single study of uncertain relevance to scientific investigation. The disdain of philosophers has been an appropriate reaction to such efforts.

Further, there is good reason to exercise caution before revising prescriptive programs for science. The scientific method has been tremendously successful for achieving understanding, prediction, and control. Just as one tries to preserve successful scientific theories, in part because they are far easier to discredit than create, there are grounds for even more stubborn attempts to preserve broader "paradigms" or philosophical systems, especially those that have been remarkably successful. The philosophy of science and the scientific method can be viewed as just this, a broad paradigm or world view. Given its power and success, and given the difficulty of altering or reconstructing it to good effect, a prudent approach to revision is advisable.

It is *not* clear, however, why or how science has been so successful. We have no description that can adequately account for this success. Further, it is doubtful that the success of science can be attributed to the strict application of philosophical prescriptions or programs. Not only have these philosophical programs changed over time, nor have mutually exclusive prescriptive programs coexisted from virtually the time science originated: in addition, mounting evidence shows that what scientists actually do and can do often holds little resemblance to what philosophers believe they have done or should do. Weimer (1979) makes this point without reservation when discussing logical empiricism:

The distance from which logical empiricists view science is enormous, and their program alienates them from the problems of the actual researcher. . . . No wonder the irony of the logical "-isms" is that despite the fact that they have more thoroughly articulated a picture of science and its nature than any other group of philosophers in

history, their program is virtually useless to the practicing scientist.
(p. 17)

Feyerabend (1979) is similarly emphatic when he states that "there
is not a single interesting event in the history of science that can be
explained in the Popperian way and there is not a single attempt to
see science in perspective" (p. 93). Therefore, although there are
justifications for caution, the philosopher cannot take refuge in the
claim that the success of science establishes a special status for the
philosophy of science and renders it exempt from empirical challenge.
Those claiming that the bulk of scientific accomplishment has arisen
from philosophical programs may find themselves much like the fa-
ther who takes pride in a child he later finds out is not his own.

Actually, although many current philosophers claim otherwise,
empirical observation has always played a critical role in the construction of scientific prescriptions. In the past, the mutual dependence between science and philosophy was openly acknowledged.
Many of the great philosophers, such as Aristotle and Descartes, were
empiricists or scientists who relied heavily on systematic observations
when constructing philosophical systems. The relationship between
science and philosophy was seen as dialectical. Over time, however,
philosophers came increasingly to claim their divorce from science,
arguing that scientific findings have no bearing on philosophical assumptions. This claim is popular today; it is often illusory, however,
and no divorce has taken place.

Perhaps all of philosophy, but certainly epistemology, is partially
based on observation. Virtually all epistemological systems make
empirical claims about how individuals construct knowledge and make
use of observation. Empirical claims are certainly made in prescriptions for science, which always include assumptions about means-ends
relationships—the form of knowledge that will result if certain prescriptive programs are used. Even purely logical systems are initially
based on interactions with or observations about the world, as Piaget
(1952) has so brilliantly shown. If empirical claims are unavoidable,
then observation, in some form, can make a potentially useful contribution to the construction of epistemological systems. Are there any
grounds to claim that we should not utilize what is perhaps the most
advanced observational technique—the scientific method? Few doubt
that the scientific method has unique advantages as a tool for guiding
observation. It is thus difficult to maintain that science is not a useful

method for gathering observations that are relevant to epistemology.

Regardless of their claims, philosophers have continually utilized scientific findings. For example, ontological systems have been vastly altered in response to new scientific insights. No one espouses Aristotle's doctrine of the four humors; but in the absence of scientific findings discrediting this model, it might still be in vogue. The impact of science on epistemological systems, especially philosophical programs for science, is also obvious. The uncertainty principle and Einstein's theory of relativity have clearly played a part in the abandonment of attempts to develop a purely objective knowledge base. This is not to claim a causal or direct linear relationship between scientific discoveries and philosophical programs. Rather, the relationship has been, and will continue to be, dialectical. It is better to openly admit this relationship and to use scientific methods freely for pursuing observations relevant to systems of philosophy. The use can be indirect, as when scientists make discoveries that bear on philosophical assumptions; or directly, as when they try to design tests of philosophical assumptions. The psychology of science can provide indirect and direct tests of philosophical assumptions about the construction of scientific knowledge by virtue of its unique suitability to the study of cognition. These tests are potentially relevant not only to descriptions of science, but to prescriptions for science as well.

UNDERLYING ASSUMPTIONS AND THEIR RELATIONSHIP TO THE STUDY OF JUDGMENT

Basic underlying assumptions can now be summarized by a series of statements. Complex information and considerations are relevant to the great majority of scientific judgments. Acognitive descriptions of science are incomplete and lead to implausible conclusions; further, they often fail to recognize that, like data, social factors do not directly shape scientific belief. Social factors can be conceptualized as forms of input that either provide information or affect exposure or access to information. All forms of information must be interpreted, and human cognition is the basis for the interpretation or construction of information. Therefore, cognition is the final common path in the construction of scientific knowledge. The study of this final common path potentially defines the central task for the psychology of science, which is uniquely suited to the study of human cognition.

Knowledge of individual human cognition provides criteria for evaluating the plausibility of assumptions about scientific knowledge construction, assumptions that appear not only in descriptions of science but also in prescriptions for science. Philosophers of science depend, in part, on empirical observations when constructing and modifying prescriptive programs. Although there are justifiable reasons to exercise caution before modifying prescriptions for science, these reasons do not justify blanket refusal to use the scientific method when attempting to generate the observation base upon which these prescriptive assumptions are built. Psychological study of the construction of scientific knowledge may uncover findings that can aid in the development of prescriptions for science or that will render descriptive or prescriptive assumptions implausible and in need of revision.

These assumptions provide a context in which to consider the psychological study of human judgment and its relevance to science. Judgment research is uniquely important to the study of science. It is perhaps the only area of psychological study in which results are directly relevant to science and also sufficiently robust to serve as criteria against which to evaluate the plausibility of descriptions of, and prescriptions for, science. Moreover, available findings indicate that certain assumptions about the construction of knowledge that appear in descriptions and prescriptions probably do not meet the criterion of plausibility. These particular assumptions are highly relevant to science because they involve core issues. Put succinctly, judgment findings strongly suggest that many popular descriptions of science are inaccurate and that many prescriptive programs are not feasible because they require judgments scientists cannot possibly make. In the remainder of this book, I will attempt to make a case for this thesis and draw out its implications.

CHAPTER 2

Human Judgment Abilities

THE DOMAIN OF INQUIRY

Few topics in psychology have received as much attention in recent years as the study of human judgment. Most studies share certain characteristics: Individuals have access to or are directly provided with information. Investigators then study the interpretation that is made of the data or the process by which these interpretations are reached. When interpretation is of interest, one studies this phenomenon by reference to the specific conclusions reached about the information or the actions taken or decisions made following exposure to the information—for example, the prediction of some criterion or the selection from among alternatives of a specific course of action. When ongoing behavior or process is studied, strategies of problem solving are most frequently of central interest. Information, then, is usually the independent variable, and some ongoing behavior or specific response (which is related to the information) is the dependent variable.

Although most judgment studies do have common features, there is wide variation among the specific research topics. For example, investigators have studied judgments about causality, frequency, and covariation. Kelley (1972) investigated individuals' judgments about the causes of another's behavior—whether they attributed the behavior to the actor's intentions or to situational factors. Tversky and

Kahneman (1973) studied individuals' assessments of the frequency of events, finding that these judgments were significantly influenced by the salience of specific occurrences. A more salient but less frequent event was commonly believed to occur more often than a less salient but actually more frequent event. Ward and Jenkins (1965) analyzed the assessment of covariation, or the relationship between two events or occurrences, and found that individuals often fail to consider most of the information needed for formulating accurate judgments. Others have investigated prediction and the formulation, maintenance, and reformulation of hypotheses. Tversky and Kahneman (1978) studied the impact of base rate information on predictions. Base rates can be thought of as the frequency with which events occur. Subjects were provided with base rates for suicide among single and married individuals and information about the proportion of single and married individuals in the population; they were then asked to judge the likelihood that a suicide victim was single. Tversky and Kahneman found that manipulations of base rate information generally had little or no impact on assumed likelihood. Lord, Ross, and Lepper (1979) demonstrated that individuals maintained a prior hypothesis, even when presented with conflicting evidence, and that subjects reinterpreted ambiguous data as supportive of their hypothesis.

In studies of these varying forms of judgment, individuals may be presented with information about either the social or physical domain. Judgments about social information require the evaluation of either the actions or behaviors of persons. As an example, one might be asked to judge the likelihood that another, or themselves, would behave aggressively when faced with a certain set of circumstances. Judgments about physical information require the evaluation of objects or the evaluation of events that do not involve human interaction, such as whether the presence of symptom "x" indicates that a disease is present or whether cloud dusting is associated with more frequent rainfall.

Given the aims of this book, no purpose would be served by attempting to provide comprehensive coverage of this broad range of studies. (The interested reader is referred to Nisbett and Ross's 1980 book.) Rather, we will consider the judgment research that has special relevance to science. In contrast to other treatments of the judgment literature, this coverage is not an end in and of itself. It is rather an initial step toward consideration of the characteristics and limitations of human judgment and their relevance to the scientific endeavor.

As one might expect from the discussion in chapter 1, I will emphasize the study of judgment under conditions in which complex information is presented to the individual. Complexity is not only a product of the amount of information; among other things, it is also dependent on the number of relevant dimensions or parameters present in the data and on the interactions and patterns of relationship that exist among these dimensions. For example, interpreting fifty pieces of information all bearing on one dimension rarely presents as difficult a judgment task as interpreting five pieces of information about five dimensions that are interrelated in a complex manner.

TYPICAL LEVEL OF PERFORMANCE ON JUDGMENT TASKS

How noble in reason? The study of human judgment may be transforming this long-accepted statement into an open question. Goldberg (1959) conducted one of the earlier investigations of human judgment performance. He studied the ability of judges to use the Bender Gestalt test to differentiate patients with and without organic brain disease. The Bender Gestalt consists of nine geometrical designs, which the patient attempts to copy or reproduce as accurately as possible. The test was among the most widely used psychodiagnostic instruments at the time of Goldberg's study, and it remains equally popular today because of its assumed sensitivity to organic brain dysfunction. In Goldberg's study, the records of patients who had completed Bender Gestalts were randomly selected from the files of a Veterans Administration hospital. The first fifteen patients that had been previously diagnosed as suffering from organic brain damage were assigned to one group. The first fifteen patients that had been diagnosed as suffering from a psychiatric illness, but not from any form of organic brain damage, were assigned to another group. Special care was taken to ensure that patients in this latter group were free of organic brain disease. Patients were included only if their psychiatric diagnosis was clearly agreed upon, they exhibited no symptoms that might be associated with organic brain dysfunction, and they had no history of head trauma.

Only one of the three groups of judges that evaluated the Benders consisted of professionals, and it is the performance of the professional group that is of interest here. The group members were four Ph.D. psychologists, with four to nine years of experience using the test. The psychologists were given face sheets and asked to sort each Bender into either an organic or nonorganic group. The Benders were divided into three sets of ten protocols, with different frequencies of organic and nonorganic patients appearing in each set. This allowed Goldberg to assess the relationship between these frequencies and the diagnostic hit rate, thus providing a check on the possibility that the 50 percent base rate of organic patients in the total sample, a rate substantially higher than that found in most clinical populations, would result in spurious findings. There was no relationship, however, between the number of organic patients (which ranged from two to eight) in the three sets of ten protocols and diagnostic accuracy, and results obtained across the three sets can be grouped together.

The findings were somewhat alarming. Given the conditions of the study, random guesses would result in accurate identifications about 50 percent of the time, or about a 50 percent hit rate. The mean hit rate for the four clinicians did surpass this level, but only marginally, falling at 65 percent. To place this finding in perspective, it should be recognized that in comparison to Goldberg's patient sample, Bender interpretation is probably more difficult with most clinical populations. Few patients exhibit the extreme forms of disturbance characterizing the patients in Goldberg's study. Further, most clinical populations contain far fewer organic patients, which likely leads to a substantially higher frequency of false positive errors (calling nonorganic patients organic) than that exhibited by these four psychologists. In other words, the clinicians' hit rate in this study is probably higher than clinicians' hit rates in most other settings. Therefore, the judgment standard achieved in Goldberg's study is far from satisfactory and hardly a basis for confidence.

Hoffman, Slovic, and Rorer (1968) investigated the ability of radiologists to establish a differential diagnosis of obvious relevance, that of benign versus malignant gastric ulcers. After consulting with a gastroenterologist, the experimenters derived seven major X-ray signs considered central to the diagnosis of cancer. Nine radiologists, six established professionals and three medical residents, were asked to evaluate 192 cases. A total of 96 different cases appeared within this sample, with each case appearing twice, thereby providing information about intrasubject reliability. The cases were ordered randomly, and each was rated on a scale of 1 to 7, ranging from definitely benign to definitely malignant. The median correlation among pairs of radiologists was only .38. The correlations for the judgments of three pairs of radiologists were *negative*. Intrasubject test-retest reliabilities fell as low as .60. Although there was no direct information about the validity of the radiologists' judgments—because criterion information was not available—it is a basic measurement principle that reliability is a necessary condition for validity. Given the low reliability of the judgments, the implication is clear. Incidentally, unknown to the radiologists, the cases were hypothetical. There is no reason to assume that results would have been better with actual cases, although the practical consequences may have been of life and death significance.

Oskamp's (1965) study provides another demonstration. In his investigation, detailed information about an individual was presented in the form of written material to thirty-two judges; eight were clinical psychologists, and five of these eight held Ph.D. degrees. The judges were then provided with a series of multiple-choice items with five alternatives. For each item, a situation was described; the judges then guessed which of the five multiple-choice alternatives, each describing a response to the situation, had occurred. All of the situations had actually taken place and the individual's responses were known, thereby providing an objective basis for determining the accuracy of the respondents' answers. There were no significant differences in the level of accuracy obtained among the groups of judges when they were separated by level of training. For example, the accuracy of the Ph.D. psychologists did not surpass the accuracy of those without Ph.Ds. Further, the accuracy of all groups of judges was surprisingly low, so low in fact (mean percentage correct was 28 percent) that the percentage of correct responses was not significantly greater than would be that expected by chance alone.

Physicians and psychologists are not the only judgment experts who have demonstrated their limitations. Studies of judgment have cut across many fields and situations, with a consistency of findings in the face of widely varying domains that builds a powerful case. For example, Horvath's (1977) study covers an area that diverges from those covered in the studies mentioned above, although it is similar to the investigation of Hoffman, Slovic, and Rorer in two respects: it involves a judgment of major practical importance and one that is often assumed to be performed accurately. Horvath studied the capability of polygraph operators to differentiate deceptive and veridical reports. Investigations of lie detection often introduce confounds that make interpretation of results problematic (Lykken 1979), and Horvath's study is one of the few that assess judgment accuracy under the conditions in which the polygraph is actually administered. In this field study, ten trained polygraph operators independently evaluated polygraph records selected from a police department file and decided whether or not the subject was lying. All subjects had denied committing a crime when polygraph testing was conducted. Fifty-six records were evaluated, twenty-eight from suspects who had later been shown to be innocent by the confession of another individual and twenty-eight from suspects who had later confessed to their guilt. Established innocence or guilt was used as the criterion for evaluating the operators' hit rates. The results were anything but impressive. On the average, almost 50 percent of the truthful subjects were incorrectly judged to be lying, a level that could be achieved by making random guesses. As was the case in Goldberg's and Oskamp's studies, the overall hit rate exceeded chance only slightly, falling at 64 percent.

These studies are representative of the disappointingly low level of performance found in many judgment studies (e.g., Einhorn 1972; Goldberg 1959, 1968b; Slovic & MacPhillamy 1974). But is this demonstrated level of performance a product of the individual or the data? What if the data do not contain the information necessary to make accurate judgments—that is, what if they are not sufficiently correlated with the criterion one is trying to predict? If this were the case, the low level of performance might reflect limits in the information, rather than limits in human judgment capacity. But a series of studies provide strong grounds to dispute any general claim that poor judgment performance is nothing but an artifact of limitations in the information.

TWO METHODS OF JUDGMENT

The relevant studies are those comparing human judgment with other methods of judgment. Paul Meehl is to be given major credit for bringing the importance of such study to the attention of a broad audience. In a clearly conceptualized and articulated work, Meehl (1954) defined terms and described various considerations that subsequently have been frequently confused. It is thus important to describe carefully the comparative judgment studies Meehl had in mind, which involve the comparison of two methods of judgment. In the first method, the human judge, using his or her head, makes a judgment based on available information. In the second method, the same information is available; but the human judge is eliminated and the information is anlayzed or combined mechanically or statistically, with "conclusions" based strictly on empirically established relationships between the information and the criterion of interest. The generic term for the first method, which applies whether the judge is doctor, psychologist, or corporate chief, is clinical judgment. This term does not indicate that the judgment is necessarily made by a clinician or in a clinical setting. The generic term for the second method is actuarial judgment.

This simple dichotomy between judgment in the head and judgment through the statistical table is presented for the sake of clarity and does not include consideration of the complications inherent in the study of clinical versus actuarial judgment. In many cases, a neat dichotomy of judgments into clinical or actuarial categories is not possible. For example, how does one classify judgment processes when the data include statements about actuarial relationships, which the judge may choose to use or disregard, or which one must combine in one's head because an actuarial procedure for combination is lacking? How does one classify when the judge, although analyzing data in his or her head, makes a decision partially or fully based on known statistical relationships between data and criterion? Clinical judgment does not, by definition, mean subjective, nonempirical, purely intuitive judgment. It is clear that most judges partially depend on knowledge of relationships between data and criterion, some or most of which has been derived by empirical techniques. For the moment, these subtleties are acknowledged so that they can be dismissed. Even when they are considered, the findings are the same (as we will see later), and to dwell on them now would only obscure central issues. For present purposes, discussion will be limited to basic comparisons of clinical and actuarial judgment as the terms have been defined above.

Comparison of the Two Methods

Einhorn's (1972) comparison of clinical and actuarial judgment affords the opportunity to examine whether poor performance could be solely attributed to limitations in the information. Three highly trained pathologists predicted survival time following the initial diagnosis of Hodgkin's disease, a form of cancer. A subset of 193 cases was selected from a large metropolitan hospital. The pathologists were provided with a biopsy slide that was assembled when the patient first reported to the hospital, but they were given no other information. The biopsy slides established the diagnosis of Hodgkin's disease. Before making predictions, the pathologists first identified nine histological characteristics relevant to determining the severity of the disease. Then, working from the biopsy slide of each patient, each histological characteristic was rated on a five-point scale of severity. The pathologists also provided, along a nine-point scale, a global rating of the overall severity of the case, which they were free to formulate in any way they chose. For all of the scales, severity was rated in ascending order, such that lower scores represented lesser severity and higher scores greater severity. Level of severity and length of survival are assumed to be negatively correlated. Therefore, one would expect a negative correlation between scale scores and length of survival.

All of the patients in the study had died at some point following the initial diagnosis, obviously providing objective information about length of survival. As one might expect from the studies described before, the pathologists' performance on this admittedly difficult task was poor. For example, the global ratings of the three judges showed respective correlations of .00, +.11, and −.14 with length of survival. The second correlation is in the positive direction, demonstrating a tendency for this judge's predictions to be inversely related to actual length of survival.

If the pathologists' poor judgment performance reflected nothing but limitations in the information, then no other method for analyzing the data should be able to surpass the level of performance they achieved.[1] Actuarial formulae, however, which were based on the judges' ratings of the nine histological characteristics, did achieve a significantly higher level of performance (see Einhorn's paper for details about the construction of the actuarial formulae). The formulae were developed using the first one hundred cases and then applied to the remaining ninety-three. Applying actuarial formulae to a sample that is independent of the one from which they were developed is known as a cross-validation procedure. For reasons that need not be detailed here, cross-validation of actuarial formulae is necessary to avoid spuriously high correlations between the predictive data (in this case the judges' ratings) and the outcome data. Under cross-validation conditions, each of the three actuarial formulae easily surpassed the judges. For example, one formula achieved a mean correlation of $-.30$ between scores and the actual length of survival. Einhorn's study, then, provides a clear case in which judges' poor performance cannot be solely attributed to limitations in the information. The judges did generate ratings of histological characteristics that had potential predictive value: it was their ratings that provided the information base for deriving the actuarial formulae. But the judges were largely unable to use this information properly. It was only when actuarial methods were used that the predictive value inherent in the information, but untapped by the judges, emerged. The judges themselves, and not the value of the information, accounted for their dismal predictive performance.

Goldberg (1965) studied the comparative accuracy of actuarial and clinical methods for assigning individuals to diagnostic categories. Psychiatrically hospitalized patients were diagnosed as neurotic or psychotic based on the information available from the Minnesota Multiphasic Personality Inventory (MMPI), a personality test they completed. The MMPI is a widely used psychodiagnostic instrument that provides scores on a series of scales, each measuring a different aspect of personality. It is generally accepted that MMPI interpretation requires one to consider not only the score on each scale, but also the interrelationships among the scores across the scales. For example, a high score on one scale may have a completely different meaning if it is the only high score, if it is only one of a number of high scores, or if there are other scores that are higher. In Goldberg's study, the accuracy rate of twenty-nine judges (thirteen psychologists and sixteen predoctoral clinical trainees) was compared with the rate obtained using an actuarial technique developed through the statistical analysis of 402 previous MMPIs. None of these MMPIs was used in the comparison study. The diagnoses of the judges and the formula were compared with the official hospital diagnosis assigned to the patient at the termination of his or her hospital stay, which served as the criterion.

Judgments were made for 861 individuals (MMPIs). No judge outperformed the actuarial formula. This was true even though the actuarial formula had been devised from a separate data base and was used under cross-validation conditions, preventing an inflated hit rate based on fortuitous statistical relationships (as often occurs with statistical methods that combine multiple variables). Two aspects of Goldberg's findings are of special interest. First, even when the judges were provided with advantages, they still could not achieve better results than those obtained with the use of the actuarial formula. For example, judges were given practice with thousands of additional cases, in which feedback was provided about judgment accuracy. This information was not made available to the actuarial program, and it probably could have been used to modify and improve this program. Second, the actuarial formula that outperformed the judges was remarkably simple (sum three scales, subtract the sum of two others, and use a cutoff point). And yet by simply combining the five scales, without attempting to assign optimal weights to each and without performing more than the most rudimentary mathematical operations, it was possible to outperform every judge.

Dawes (1971) found that even simpler actuarial techniques could outperform human judges. Dawes compared the clinical and actuarial prediction of success in graduate school. The criterion for success was faculty ratings of students attending a graduate school program; these ratings were of established reliability. Clinical predictions were made by an admissions committee, which had access to information about four variables: undergraduate institution attended, Graduate Record Examination scores, grade point average, and letters of recommendation. These first three variables, but not the last one, were used for constructing actuarial procedures. Dawes devised an actuarial formula that easily outperformed the committee, but this is the least remarkable aspect of his findings. Using the actuarial method, it was possible to eliminate 55 percent of the applicants who were considered by the admissions committee and later rejected, without eliminating a single applicant who was considered and later accepted. Dawes (1971) conservatively estimated that the use of such a formula would save $18 million annually by eliminating thousands of work-hours spent on the evaluation of candidates certain to be rejected. More important, Dawes found that it was possible to outperform the admissions committee by making actuarial predictions on the basis of a single variable (grade point average). A simpler actuarial formula is hardly possible, and yet it was still sufficient to surpass the performance of the committee.

The studies of Einhorn, Goldberg, and Dawes provide demonstrations of two findings that appear consistently in the literature comparing clinical and actuarial judgment. For one, these studies demonstrate the human judge's inability to outperform the actuarial method. In Meehl's (1965) review of fifty-one studies comparing the two methods, only one demonstrated a significant advantage for the clinical method, and the veracity of this study was strongly disputed by Goldberg (1968a). Thirty-three studies demonstrated a significant advantage for the actuarial method, with seventeen showing no significant advantage for either method but a general trend towards the superiority of actuarial methods. Meehl (conversation with the author, December 1981) recently indicated that his ongoing review of the numerous comparative studies that have been conducted since his 1965 tally has not yet uncovered a single study showing a significant advantage for the clinical method. The uniformity of the findings is remarkable, given the extreme rarity of obtaining such consistency in the social sciences. Further, in many studies (such as those of Goldberg and Dawes) extremely simple actuarial techniques were used. A second finding, then, is that simple actuarial techniques consistently equal or surpass the performance of human judges. Although more complex actuarial procedures are potentially available, which optimally weight variables and combine them in a configural manner, they are unnecessary. Linear procedures with unweighted variables are sufficient. That the least powerful actuarial procedures consistently equal or outperform judges leads one to conclude that a robust phenomenon has been uncovered.

Challenges to the Demonstrated Superiority of Actuarial Methods

The studies supporting the superiority of the actuarial method have been challenged. Many of the points of contention are based on misunderstandings of Meehl's initial statements, but some critics have raised pertinent questions requiring serious consideration and empirical analysis. One early objection was raised by McArthur (1956a, 1956b), who contended that the studies required clinicians to perform judgment tasks atypical or nonrepresentative of those encountered in their everyday work. Prediction of artificial criteria in artificial settings prevented access to the information and utilization of the knowledge base essential to expert clinical judgment. In other words, the clinician was not allowed to perform the tasks for which expertise had been acquired, thereby creating a major disadvantage.

But this contention was dispelled through subsequent study. A diversity of judgment tasks have now been investigated (Meehl 1965), which are fair representations or actual instances of the tasks conducted by a wide range of professionals. Medical diagnosis, psychiatric diagnosis, prediction of job success and job satisfaction, success in academic or military training, and recidivism and parole violation are among the judgment tasks studied. In many of these studies, such as those conducted by Einhorn (1972) and Goldberg (1965), competent professionals have performed the tasks for which they claim expertise. Even under these conditions, there has been virtually no evidence uncovered for the superiority of the clinical method. Artificial and natural circumstances alike, the results remain the same.

A second and related contention (e.g, Holt 1958) was that clinicians were not given access to data that, in real-world judgment situations, are critical for arriving at accurate clinical predictions. For example, in some studies clinicians were required to make diagnostic decisions without interviewing patients or at least reviewing interview data, which are assumed to contain information necessary for maximizing judgment accuracy. To address this and related issues, Sawyer (1966) conducted a thorough literature review and a reanalysis of the studies comparing actuarial and clinical judgment. Sawyer's article is a major contribution to the area and, even today, retains its full relevance. Before examining it in detail, let us review essential background information.

Meehl's (1954) book mainly addressed clinical versus actuarial methods of data *combination*. As Meehl unequivocally stated, fair tests of the specific comparison he had in mind require that the same information be made available to the clinician and to the actuarial formula. Some information, however, may be difficult to code into a format that permits actuarial analysis. Clinical impressions and interview data, which Holt claimed were essential to clinical prediction, often present particular difficulties. Hence, in the interest of fulfilling the conditions for a fair comparison, when such information could not be coded and made available for actuarial use it was not made available to the clinical judge either. This omission provided potential fuel for Holt's claim that lack of access to needed information placed the clinician at a disadvantage. Even more serious problems occurred when investigators provided different information to the clinician and the actuarial method, thereby violating Meehl's conditions for a fair test. In some studies, clinicians had access to information that was not provided to the actuarial method. In others, clinicians had access to clinically collected information only (e.g., interview transcripts), whereas the actuarial program had access to mechanically collected information only (e.g., scores from psychological tests). And in other studies, the method of combining information was held constant and the method of obtaining information varied. For example, comparisons were made between clinical predictions based on clinically versus mechanically gathered information, or between actuarial predictions based on clinically versus mechanically gathered information. Unfortunately, across this range of studies most researchers claimed they were comparing clinical and actuarial judgment. It was not that the comparisons, per se, were invalid or without value. It was rather that different studies had been lumped together as tests of the same question, when they actually often addressed different questions; and that many studies, regardless of claims to the contrary, were *not* valid tests of the comparison Meehl had outlined.

Sawyer sought to untangle this confusion. He began by making the critical distinction, which had often been overlooked, between methods of collecting information and methods of combining information. Meehl had placed major emphasis on comparing the predictive accuracy of clinical and actuarial methods for *combining* information, under conditions in which *neither* method was provided with an unfair advantage. Sawyer reiterated that a valid comparison of this type required that the same information, regardless of how it was collected, be made available to both the clinical and actuarial methods of data combination. Sawyer, however, argued that Meehl's emphasis on the comparison of methods for combining data potentially led researchers to overlook comparisons of methods for collecting data, which Sawyer also considered to be of major relevance. Believing that the term *actuarial* did not apply equally well to methods of collecting information, Sawyer substituted the term *mechanical*. Mechanical methods for collecting information are characterized by prespecified procedures that are *not* altered on the basis of clinical judgment. A paper-and-pencil personality measure and a standardized list of interview questions are examples. Mechanical methods for combining information are simply actuarial procedures for "interpreting" the collected information as described by Meehl (1954).[2]

Sawyer, then, distinguished two methods of collecting information

Data Combination

Data Collection	Clinical	Mechanical
Clinical	A	B
Mechanical	C	D
Both	E	F

Figure 1. Methods of collecting and combining information.

—mechanical and clinical—and two methods of combining information—mechanical and clinical. These four methods or conditions, which are represented by cells A-D in figure 1, make numerous comparisons possible. As Meehl proposed, one could compare clinical versus mechanical methods for combining information. Either clinical or mechanical methods could be used to gather this information, although the identical information would need be made available to the clinical and actuarial methods of combination.[3] For example, one could compare cells A and B, or C and D, of figure 1. Other comparisons, however, addressing different questions, could be made. One could compare clinical versus mechanical methods for collecting information, while holding the method of combining information constant (e.g., comparison of cells A and C, or B and D). Or one could vary both the method of collecting and combining information, as by comparing conditions in which clinical versus mechanical methods were used for both collecting and combining information (cells A and D).

Sawyer also presented four other possible methods or conditions, two of which are illustrated in cells E and F of figure 1.[4] Cell E represents the condition in which both clinically and mechanically collected information are available and are combined clinically. Cell F represents the condition in which both forms of collected information are available and are combined mechanically. These additional conditions obviously also make many other comparisons possible. Further, cell E satisfies Holt's requirement that the clinician be provided with the information normally available in everyday practice—both mechanically and clinically collected information.

Sawyer's taxonomic system covers the full range of conditions studied in the clinical-mechanical debate, making it possible to classify the types of comparisons that had been made. This clear conceptual scheme permitted Sawyer to analyze each type of comparison separately, circumventing the problems created when investigators lumped studies together that addressed different questions. Using this classification scheme, Sawyer reexamined the studies comparing clinical and mechanical prediction. He covered all studies ($N = 45$) that provided adequate information for such comparisons and that included more than one predictor for purposes of mechanical combination, unless it was clear that additional predictors would not change the conclusions (i.e., when even with one predictor the mechanical method was superior). The methods of judgment appearing in each study were classified into one of the categories described above, depending on the manner in which information was collected and combined. For example, if the method involved clinical collection and clinical combination, it was classified as a member of category A (refer back to cell A of figure 1). The specific comparisons made in each study were then grouped into types, and across the studies all comparisons of the same type were combined. For example, all comparisons between the method used in category A and that used in category B were grouped together. The forty-five studies provided seventy-five pairs of comparisons (some studies included more than two methods), which Sawyer grouped into twenty types.

Remarkably, not one of the seventy-five comparisons opposed any other comparison. Within each type of comparison, there were no instances in which one study found one method better while a second study found another method better. Across these comparisons, the mechanical method for combining information always equalled or surpassed the clinical method for combining information, regardless of the method used for collecting information. This finding held true even when the clinical judge was provided with both clinically and mechanically collected information and the actuarial method was provided with only one of these forms of information. It follows from this finding that in every case in which the method of *collecting* data was held constant and that of combining data varied, the me-chanical method equalled or surpassed the clinical method. This last finding offers strong support for Meehl's hypothesis regarding the comparative predictive power of clinical and actuarial methods for combining information, but it clearly contradicts Holt's hypothesis. Even when the clinician and actuary were provided with access to both mechanically and clinically collected information (cells E and F of figure 1, respectively)—the circumstance in which, Holt argued, the clinical would outperform the actuary—the clinical method of combination never surpassed the mechanical method.

This powerful demonstration of the advantage held by mechanical methods for combining data is of central importance here, although other of Sawyer's findings are relevant to later discussion. Some of the studies, for example, provided clinicians with clinically and me-chanically collected information as well as with predictions resulting from the mechanical combination of this information. The clinician was then asked to make his or her own predictions (to combine the information), with the choice of simply adhering to the mechanical prediction, using it as an additional piece of data, or ignoring it com-pletely. The study of this condition, which Sawyer labeled "clinical synthesis," addresses a critical issue: the clinician's ability to identi-fy instances in which he or she should supersede mechanical predic-tion and instead rely on clinical prediction. In many everyday judg-ment situations, the individual must decide whether to rely on the mechanical method, even when it might not consider or adequately weigh important variables and, therefore, might lead to a less accurate prediction than the clinical method. If clinicians can correctly identi-fy such instances, they will have an advantage over the mechanical method.

Meehl (1954) has coyly illustrated this possible circumstance with the "broken leg" example. A professor reliably attends movies; how-ever, on one occasion he breaks his leg before setting off for the mov-ies. Mechanical predictions often do not take into account unexpected or low-frequency occurrences. Therefore, the mechanical method would probably lead to the prediction that this incapacitated profes-sor would be in attendance. In this case, one would be better off superseding the mechanical prediction and instead making the obvious clinical prediction that the professor would miss his movie. If one could find actual instances of broken leg examples, or other such circumstances in which the clinician could successfully determine when to supersede mechanical predictions, it would at least support the argument that clinicians are able to decide when and when not to rely on mechanical prediction.

Sawyer, however, found no broken leg examples among the studies comparing clinical synthesis to other methods of judgment. In no case did clinicians, given access to clinically and mechanically collected information and to the results of the mechanical combination of this information, outperform the mechanical method for combining information. Clinicians, then, could not properly decide when to supersede mechanical predictions and would have been better off uniformly adhering to these predictions. Sawyer had access to only three studies; however, according to Meehl, subsequent research has not challenged this finding (conversation with the author, December 1981). To date, predictions based on clinical synthesis have never surpassed predictions based on mechanical methods alone. This finding raises serious doubt about the ability of individuals to identify instances in which mechanical predictions are less accurate than clinical predictions and to use this knowledge to improve upon mechanical predictions. There is no evidence, then, that we can rely on our own judgment about when to rely on our own judgment.

Sawyer also found that Holt was partially correct but that this partial accuracy only served to undermine Holt's contention further. Holt had claimed that clinically collected information increased predictive accuracy. Sawyer did find that predictive accuracy increased when clinically collected information was added to mechanically collected information, but this did *not* occur when the information was *combined clinically*. In fact, when clinically collected information was added to mechanically collected information and then combined clinically, judgment accuracy often decreased. Thus, clinical combination often resulted in higher predictive accuracy when only mechanically collected information was available. In no study did the addition of clinically collected information improve *clinical* prediction. In contrast, when mechanical methods of combination were used, the addition of clinically collected information to mechanically collected information did, at times, increase predictive accuracy.

Clinically gathered information, then, could be useful, but only under circumstances directly opposing those hypothesized by Holt. Holt had argued that, if clinical judges were given access to clinically collected data, the discrepancy between the clinical and mechanical methods of prediction would be narrowed or even reversed. But when this information *was* provided, the performance of the clinician either remained the same or declined, while the performance of the mechanical method frequently improved—the gap between the clinical and mechanical methods of prediction only widened. Sawyer's review strongly suggests that clinicians do collect valuable information but that it only increases predictive accuracy beyond the level achieved using mechanical information alone when it is combined mechanically. Therefore, the clinician might best contribute to predictive accuracy by providing clinically collected information, which is then combined mechanically, and not by attempting to integrate this and other information.

Finally, Sawyer uncovered an intriguing discrepancy between optimal and actual judgment practices. When Sawyer's review was published, it was already clear that the numerous studies demonstrating the inferiority of clinical prediction had done little to alter the general preference (at least among psychologists) for clinical judgment methods. Sawyer found that when clinicians did use mechanical methods for combining information, it was most often in situations in which relatively simple information was available. Integration of complex information was still left to the clinical judge. Ironically, it was under this latter condition that the mechanical method showed the greatest advantage over the clinical method.

The failure to make optimal use of mechanical methods for combining information may also characterize current practices in science. On lower-level tasks, mechanical methods of combining information, such as statistical tests, are common. But, assuming that the new view of science has any credibility, it is for the most complex reasoning tasks, such as theory evaluation, that scientists depend less on systematic or mechanical methods for integrating information and more on clinical judgment. Actually, few systematic methods for performing higher-level tasks are currently available, and scientists usually *must* rely on clinical judgment. Paradoxically, it may be that these methods are unavailable for the very judgments that scientists perform least capably and where mechanical methods are most needed.

SUMMARY

Judgment studies have consistently demonstrated that the absolute level of human performance on a wide range of tasks is surprisingly low. Even the performance of experts on tasks for which they claim or are assumed to hold competence is frequently substandard and, at times, barely exceeds the level that would be expected by chance alone. The studies comparing mechanical versus clinical methods provide clear evidence that the poor performance of judges cannot be attributed solely to limits in the predictive value of the available information. In many cases, the information has far greater predictive power than can be garnered by the clinical judge, as evidenced by the significant gain in predictive accuracy that is achieved when actuarial methods are applied to the same information. Additionally, in every study published thus far (with one questionable exception), even simple actuarial methods have equalled or surpassed clinical judges. Earlier contentions, such as McArthur's claim that studies handicapped the clinician by requiring unfamiliar predictions, or Holt's claim that studies placed the clinician at a disadvantage by precluding access to critical information, have subsequently been resoundingly discredited. At this point, unless proven otherwise, one can assume that human judgment performance is often substandard and can be equalled or surpassed by actuarial methods.

CHAPTER 3

Factors Underlying
the Judgment Findings

SOURCES OF JUDGMENT ERRORS

For several years we have been able to describe experimental results
demonstrating poor performance on judgment tasks, as well as those
demonstrating the superiority of actuarial methods. But only more
recently have we been able to move beyond pure speculation about
the factors underlying these results and towards a beginning scientific
understanding. As the evidence accumulates it shows more clearly
that many of the early speculations were wrong, including the frequent speculation that bias, as it is commonly viewed, underlies most
judgment failures. As Dawes (1976) points out, intellectuals from
Plato to Freud have viewed bias as contamination, blockage, or inter-
ference of higher intellectual processes by the lower-level human
drives or processes, such as animal instincts or emotion. Bias as so
defined has been blamed for human thinking failures for centuries.
We now have compelling evidence, however, that a different type of
bias underlies many judgment failures, and that other judgment failures stem from factors that cannot at all be considered forms of bias.
For simplification, I will separate the sources of judgment error that
differ from emotional bias, as it is typically defined, into two categories. The first covers instances of bad judgment habits and failures
to use normative judgment methods; these could be considered forms
of bias if the definition is stretched. The second category, to which

the term bias simply does not apply, covers human cognitive limitations.

CATEGORY I: BAD HABITS AND FAILURE TO
USE NORMATIVE JUDGMENT METHODS

Individuals make certain judgment errors when they use judgment rules or methods that violate normative guidelines or when they fail to use available normative judgment methods. Numerous examples of judgment errors resulting from these sources have been described in the literature. For example, without recognizing it, individuals apparently often form judgments about the probability of an event based on the ease with which they can recall instances of the event. Tversky and Kahneman's study (1973) provides a demonstration of this bad judgment habit. Individuals were presented with lists of males and females. In one condition, the men appearing in the lists were relatively more famous than the females appearing in the lists, and in the other condition the relative fame of the men and women was reversed. Individuals consistently overestimated the proportion of males in the lists containing relatively more famous males and overestimated the proportion of females in the lists containing relatively more famous females, apparently because famous names could be more readily recalled.

Another example of a bad habit is the tendency to believe that consistent data have greater predictive value than less consistent data, even if the former are highly redundant (nonindependent) and the latter minimally redundant. In one experiment (Tversky & Kahneman 1974), individuals expressed more confidence in predicting a student's final grade point average when the first-year record contained all Bs than when it included many As and Cs. As a third example, when individuals make predictions, they often heavily weight content and all but ignore the predictive value of information. The research of Tversky and Kahneman (1974) again provides a demonstration. Individuals read a paragraph describing the quality of a student teacher's performance on one lesson, and they were then asked to predict that teacher's eventual success. Individuals realized it was unlikely one could predict eventual success on the basis of such limited information. Nevertheless, their predictions corresponded exactly with their evaluation of the student teacher's one performance, with no attempt made to adjust predictions away from the extremes. This strategy clearly violates a basic judgment principle, which dictates that the extremeness and range of predictions should be controlled by the reliability and validity of the predictive information.

As a final example, individuals frequently fail to recognize instances in which regression towards the mean probably accounts for occurrences and instead invent causal explanations. For example (Kahneman & Tversky 1973), pilot training instructors noted that when they praised an exceptionally good landing the subsequent landing was often poorer, and when they criticized a poor landing the subsequent landing was often better. The instructors concluded that praise diminishes performance and that criticism improves it, but it is far more likely that the observed pattern was simply a manifestation of regression towards the mean. In most areas, an excellent performance is usually followed by a less excellent performance, and a very poor performance is usually followed by a less poor performance (regardless of verbal praise or criticism). Extremes tend to regress towards the mean.

The judgment literature contains many more examples of errors stemming from bad habits and failures to use normative methods. (For an excellent review, the reader is referred to the 1974 article by Tversky and Kahneman.) Given present purposes, there is no need to list all of the examples that have been documented. Rather, the aim here is to provide a general awareness of these sources of judgment error and their relevance to science. Therefore, three such sources, which commonly result in scientific judgment errors, will be explored in detail: underutilization of base rates, belief in the "law of small numbers," and reliance on confirmatory strategies.

Underutilization of Base Rate Information

One of my colleagues, who is a fervent adherent of a psychodiagnostic instrument that has come under frequent criticism, announced the results of a recent study that "unequivocally" demonstrated the usefulness of the test in an applied situation of major importance. In this study, a series of psychiatric patients, some of whom later committed suicide, had completed the test. Based on the test results of those who had committed suicide, a series of suicidal indicators were derived. The entire sample was then reanalyzed, and it was found that all patients who later committed suicide scored above a certain level when the suicide indicators were summed. On this basis it was possible to derive a cutoff score that, in retrospect, identified all future suicide vicitms but that made few false identifications of individuals who later did not commit suicide.

On the basis of these results, assuming the study is valid, has the instrument been demonstrated to be of unequivocal value in accurately identifying those at risk of taking their lives and those not at risk? By this colleague's report, there was about a 5 percent false positive rate; that is, about 5 percent of the patients in the study were incorrectly identified as future suicides. Taking a hypothetical series of one thousand consecutive patients in an outpatient clinic and making a series of very generous assumptions, assume that the false positive rate would remain at 5 percent and assume a high suicide rate in this population, say 1 percent. Further, assume all future suicide victims were correctly identified. From the population of one thousand, 5 percent or fifty patients would be falsely identified as future suicides and 1 percent or ten patients would be correctly identified as future suicides, an error rate 5 times greater than the hit rate. If one were to use the instrument in a setting with a lower suicide rate, say one per thousand, the error rate would exceed the hit rate by 50 to 1. Questions of utility aside, what first appeared to be an almost infallible means for identifying future suicides is seen under more critical analysis to produce far more errors than correct identifications.

My colleague's unfortunate misappraisal stemmed from a failure to weight base rate information carefully, a common but, nonetheless, bad judgment habit. Consideration of base rates—the relative frequency at which events occur—is critical for optimizing judgment accuracy in numerous situations. The above example illustrates one such situation: With low frequency events, such as suicide, highly accurate information is required to increase the hit rate beyond that which would be obtained using base rates alone. Individuals frequently fail to consider, or to sufficiently weight, base rate information (Kahneman & Tversky 1973; Meehl & Rosen 1955; Nisbett, Borgida & Crandall 1976). With psychodiagnostic tests, for example, if the frequency of false positive identifications exceeds the frequency of the event to be identified or predicted, the use of the test will decrease the hit rate relative to that which would be obtained if predictions were founded on base rates alone. If the test incorrectly "says" a condition is present more often than the condition actually occurs, the test cannot increase diagnostic accuracy. But, as Meehl and Rosen point out, even when this principle applies psychologists often still use psycho-diagnostic tests instead of base rates.

Kahneman and Tversky (1973) uncovered additional instances in which individuals underutilize base rate information. Perhaps most striking was a study requiring subjects to guess the occupation (engineer or lawyer) of a person, based on extremely ambiguous personality descriptions, such as the following:

Dick is a 30-year-old man. He is married with no children. A man of high ability and high motivation, he promises to be quite successful in his field. He is well liked by his colleagues. (p. 242)

In one condition, subjects were told that 70 percent of the described persons were lawyers and that 30 percent were engineers, and in a second condition the frequencies were reversed for the two occupations. In either condition, one could achieve a hit rate of 70 percent by using base rates alone—by predicting that all the persons were lawyers when one was told that 70 percent of the sample were law-yers. Nevertheless, participants depended strongly on these almost useless personality descriptions when guessing occupation, and they were completely unaffected by manipulations in the base rates. This behavior occurred even though many of the participants recognized that the personality descriptions provided virtually useless informa-tion for determining occupation.

Many investigations, such as those cited above, demonstrate the frequent underutilization of base rate information. It is highly likely that scientists also underutilize base rate information at times and thereby decrease their judgment accuracy. The possible generaliza-tion of this and other types of judgment error to scientists will be detailed in chapter 4.

Belief in the Law of Small Numbers

A researcher working with a sample size of twenty obtains a signifi-cant finding ($z = 2.23, p < .05$, two-tailed). Assuming this significant finding accurately reflects the true state of nature, what is the proba-bility of obtaining a significant result if the study is replicated with a sample size of ten? The probability is .48, as calculated by Tversky and Kahneman (1971), who originally presented this problem. If readers overestimated this figure, they are in good company. When the problem was presented to a group of skilled experimental methodologists, most also overestimated the probability of obtaining a significant result (Tversky & Kahneman 1971). Tversky and Kahneman argue that the difficulty encountered on this problem illustrates a particular source of judgment error, which they refer to as "belief in the law of small numbers." Tversky and Kahneman claim that individuals have unrealistic expectations about the properties of small samples. It is commonly assumed that small samples are as representative, or nearly so, as larger samples, and that they conform to the laws of chance applicable to larger samples.

Belief in the law of small numbers, according to Tversky and Kahneman, is based on the erroneous conviction that the degree of similarity between the characteristics of the sample and the assumed characteristics of the population is a reliable indicator of the sample's validity. For example, if individuals believe that events in the population (e.g., coin tosses) are randomly distributed, then even small samples depicting events that appear more random than ordered are seen as more likely to have been drawn from the population. In one experiment (Kahneman & Tversky 1972), individuals believed that the sequence of coin tosses HTTHTH was more likely than either HHHHTH or HHHTTT, although the probability of the three sequences is equal. Judgments about the validity of the sample, then, are significantly influenced by the extent to which the characteristics of the sample are assumed to be representative of the population. Such a reasoning strategy creates a tendency to overlook sample size. This parameter does not reflect a characteristic common to sample and population and, therefore, does not enter into judgments of similarity or representativeness. As a result, sample size is often ignored, leading to judgments that treat small samples as if they had the same properties as large samples.

Tversky and Kahneman (1971) present numerous examples in which belief in the law of small numbers leads to judgment errors. Especially relevant are demonstrations of scientists' susceptibility to such errors, which will be described in chapter 4. In one study of nonscientists (Kahneman & Tversky 1972), subjects estimated the probability of obtaining various distributions of newborn males and females in samples of 10, 100, and 1,000 births. For the three samples, subjects were asked to guess the percentage of days the number of boys would comprise 5, 5-15, 15-25, . . . , 85-95, and 95-100 percent of the total population of births. Obviously, the larger the sample size, the less the chance of a significant deviation from the mean. For this reason, as sample size increases, extreme percentages should occur less frequently. For example, although the probability that a sample of ten will contain 70 percent boys is approximately .25, the probability that a sample of 1,000 will contain this percentage of boys is infinitesimal. Nevertheless, individuals produced virtually identical frequency distributions for the three sample sizes. Whatever the sample size, subjects assumed that the degree of similarity between the sample and the population would be equal, an example of erroneous judgments resulting from belief in the law of small numbers.

In another demonstration, Stanford University undergraduates were presented with the following problem (Kahneman & Tversky 1972):

A certain town is served by two hospitals. In the larger hospital about 45 babies are born each day, and in the smaller hospital about 15 babies are born each day. As you know, about 50% of all babies are boys. The exact percentage of baby boys, however, varies from day to day. Sometimes it may be higher than 50%, sometimes lower. For a period of 1 year, each hospital recorded the days on which more than 60% of the babies born were boys. Which hospital do you think recorded more such days? (p. 443)

The students selected one of the three possible answers: "the larger hospital," "the smaller hospital," and "about the same (that is, within 5 percent of each other)." Again, because larger samples are less likely to deviate from the mean, the larger hospital is far less likely to have recorded more days in which 60 percent or more newborns were boys. Nevertheless, over half of the students answered that the two hospitals would have about an equal number of days. The remaining answers were evenly split between the larger and smaller hospitals.

Kahneman and Tversky (1972) concluded, on the basis of this and related findings, that "the notion that sampling variance decreases in proportion to sample size is apparently not part of man's repertoire of intuitions" (p. 444). The failure to appreciate this basic principle can lead to two basic types of judgment errors. One is the false negative error, the conclusion that no relationship exists between variables that are actually related; the other is the false positive error, the conclusion that a significant relationship exists between variables that are actually unrelated. Both types of errors stem from the failure to recognize when occurrences are an artifact of small sample size. These two types of judgment error will be discussed in greater detail in chapter 4.

Confirmation Strategies

Overreliance on confirmation strategies is a third judgment habit that can lead to errors. When testing hypotheses, individuals seem to rely heavily on confirmation strategies or to concentrate their attention on confirmatory evidence. Disconfirmatory strategies are rarely used, and when disconfirmatory evidence is obtained it is often devalued or completely overlooked. (Evidence "confirming" these statements is reviewed in Nisbett and Ross [1980, chap. 8]; and Tweney, Doherty, and Mynatt [1981, part IV].) For years, informed persons have warned us of the dangers of confirmatory strategies, with Popper being perhaps the most eminent and persuasive. But the relative merit of confirmatory and disconfirmatory strategies is anything but a dead issue. For example, Kuhn (1970) has argued that closed-mindedness, which seems closely linked to confirmation strategies, makes the wheels of normal science turn. Lakatos (1970) also raised the possibility that science can progress in the absence of refutation. The logic of disconfirmation may be impeccable; science is not a purely logical pursuit, however, and it is often difficult to defend arguments for uniform adherence to more logically adequate strategies. Philosophical arguments alone will never fully resolve questions about the merits of confirmatory and disconfirmatory strategies. Instead, we need in-depth study of the strategies and the specific contexts in which they may or may not be functional.

Judgment research may provide a beginning glimpse of specific situations in which confirmation strategies are almost certainly dysfunctional. In these situations, confirmation strategies often lead to the maintenance of hypotheses in the face of sound disconfirming evidence and to marked distortions in the interpretation of data. Although these demonstrations do not provide logical proof against confirmatory strategies, they may, within the contexts studied, accurately describe likely consequences that must be accounted for by individuals taking proconfirmatory positions. For example, in his argument for the merits of closed-mindedness or confirmatory strategies, Kuhn may have to allow that they can cause major errors in data interpretation. Whether such an admission necessitates reformulation of his or other proconfirmatory arguments is an open issue.

The tendency of nonscientists to rely on confirmatory strategies and to overlook disconfirmatory evidence has been documented repeatedly. Mynatt, Doherty, and Tweney (1978) designed an artificial "research environment" in which individuals could observe and partially control a computer screen depicting moving particles. Individuals attempted to uncover the "laws" of particle motion that had been programmed into the computer. They could conduct tests of the hypotheses they formed by exercising control over the display screen. In the great majority of cases, individuals used confirmatory strategies to test their hypotheses. Failure to find support for a hypothesis often did not result in its permanent abandonment, even when results provided conclusive disconfirmation. Snyder (1978) studied subjects' strategies for testing hypotheses about another individual's level of introversion or extroversion. Snyder's experiments must be interpreted conservatively, given the possibility that salient demand characteristics may have unfairly influenced his subjects' responses. Nevertheless, the studies are impressive in demonstrating the consistency with which individuals rely on confirmatory strategies across widely varying conditions. Tversky (1977) found that individuals searched for common, rather than disjunctive, features when judging similarity, a variant of the confirmatory approach. In Wason and Johnson-Laird's (1972) study of concept learning, individuals almost always looked for instances that verified propositions and almost never looked for instances that falsified them.

Although one could provide a long list of studies showing reliance on confirmatory strategies, what evidence is there that such practices lead to judgment errors? Chapman and Chapman's (1967, 1969) research on prior hypotheses probably offers the most impressive demonstrations of judgment errors resulting from confirmatory strategies. In their 1967 study, Chapman and Chapman utilized the Draw-a-Person test (DAP), a popular psychodiagnostic aid. The test is simple. The patient draws a picture of an individual, or separate pictures of individuals of each gender, and the clinician then attempts to analyze the characteristics of the pictures and to draw inferences about the "artist's" (patient's) personality. Many clinicians share a set of assumptions about the personality traits that can be inferred from specific drawing characteristics. For example, it is assumed that drawings featuring large and accented eyes are associated with paranoid traits. Special emphasis given to the mouth is supposedly indicative of strong dependency needs. Although attempts to validate these and other assumed correlations have repeatedly failed, this research has done little to shake many clinicians' faith in the measure. The Chapmans sought to uncover why intelligent and highly trained clinicians could maintain this faith, even in the face of abundant disconfirmatory evidence.

Preliminary investigation showed that the clinicians' assumptions about drawing characteristics and personality were not unique. Similar assumptions were also held by naive observers, in this case, college undergraduates. This led the Chapmans to hypothesize that these assumptions were not founded on clinical data, but rather on common verbal associations linking particular drawing characteristics with symptoms. For example, vigilant eyes and suspiciousness are often linked in our language. Having established the presence of these prior associations between characteristics and symptoms, or what could be considered prior hypotheses, the Chapmans examined the extent to which they would be altered by disconfirming evidence. The researchers collected forty-five drawings that had been produced by patients at a state hospital. Six symptom statements were constructed, each of which described a personality characteristic that was frequently inferred from figure drawings. A single statement was *randomly* paired with each picture, ensuring that no symptom statement and drawing characteristic were systematically related to each other. For example, drawings of figures with large eyes were paired equally often with the symptom statement indicating suspiciousness as they were with each of the symptom statements indicating other personality traits.

Undergraduate students examined each pair of drawings and personality statements and were asked to report whether certain drawing characteristics were more frequently associated than others with certain symptoms. Almost all students reported systematic relationships between drawing characteristics and symptoms. Most of these relationships duplicated the relationships that clinicians generally assumed existed between drawing characteristics and personality.

The students probably held these assumptions before they examined the drawings, given Chapman and Chapman's previously described finding—that naive subjects and clinicians share similar assumptions about drawing characteristics and personality. When the students did examine the drawings, the relationships they thought they found were apparently the ones they previously believed existed. In other words, prior belief shaped the interpretation of evidence and led the students falsely to assume that this nonsupportive data confirmed their hypotheses. Chapman and Chapman argue that prior assumptions often lead individuals to maintain belief in nonexistent relationships or "illusory correlations," even in the face of disconfirming data.

If prior beliefs can lead to the conclusion that nonsupportive data provide evidence for illusory correlations, how might prior beliefs affect the detection of valid relationships that *are* supported by the data? For example, if individuals are exposed to data containing systematic relationships between variables, will previously formed illusory beliefs hinder the detection of these valid relationships? This is the central question Chapman and Chapman (1969) addressed in an extension of their DAP study. The DAP was replaced by the Rorschach, which does show some systematic relationship between signs and symptoms. The Chapmans constructed thirty Rorschach responses, covering five categories, which were modeled after actual statements by patients. Some responses, for example, referred to food, as in "It looks like an ice-cream cone." The thirty responses included six different items pertaining to each of the five categories.

One condition in this study was essentially a replication of the DAP study. Subjects were shown thirty Rorschach cards, with one response paired with two symptom statements on each card. The subjects were told that each response was produced by a different patient and were asked to report if they observed any relationships between patients' responses and symptoms. Types of responses and symptom statements were paired randomly. Again, individuals assumed that this data provided support for certain systematic relationships between responses (signs) and symptoms, and it appeared that these perceived relationships coincided with beliefs held prior to the experiment. As in the DAP study, it was not that the individuals had prior experience with the diagnostic instrument but that they had previously formed semantic associations between the verbal content of responses and personality characteristics. For example, bats and morbid thoughts are often linked together. Therefore, if a Rorschach response contains such verbal content as "It looks like a bat," individuals naturally assume that the respondent has morbid preoccupations.

In another condition, the Chapmans selected two categories of responses that independent research had shown to be actually related to certain personality traits, and they paired these responses with valid personality statements at a greater than chance level. These valid relationships were not ones that individuals intuitively assumed existed. Therefore, subjects did not enter the experiment with the prior belief that such responses and symptom statements were related to each other. Three separate groups of individuals were exposed to separate sets of Rorschach cards, in which the valid pairings occurred at a 67, 83, and 100 percent level of frequency, respectively. The other categories of responses remained randomly paired with symptom statements. All three groups of individuals believed that the data provided support for the same illusory correlations reported by individuals in the previously described condition, with this outcome, again, apparently a product of prior belief. Belief in these illusory correlations, then, was impervious to the contrary influence of the valid relationships present in this data. Of equal importance, the valid relationships were not detected or reported except in the 100 percent condition. When the valid pairings occurred uniformly, some individuals reported observing a slight relationship between the response categories and symptom statements.

The Chapmans have argued (1969) that it is not task difficulty that accounts for these results, but rather the presence of prior beliefs in illusory correlations. They presented the same essential task to another group of individuals but selected response categories for which prior beliefs were not held. Two of these categories were systematically paired with symptom statements, whereas the other four categories were randomly paired with symptom statements. Under these conditions, most individuals were able correctly to identify the systematic relationships between responses and symptoms. Apparently, then, when prior beliefs exist, there is a tendency to use confirmation strategies and to discount disconfirming instances. Distorted interpretations of the data may result. In the absence of prior hypotheses, confirmatory strategies may still be used, but freedom from prior illusory beliefs gives one a greater chance to form correct hypotheses initially.

Taken together, the findings of the Chapmans and others suggest that individuals frequently use confirmation strategies but rarely use disconfirmation strategies. Perhaps, as Kahneman and Tversky (1972) found when studying awareness of sampling principles, disconfirmatory strategies are not part of the human repertoire of intuitions. Prior hypotheses and the associated use of confirmatory strategies also potentially lead to belief in illusory correlations. Once hypotheses are formed, subsequent data are often interpreted as supportive of one's hypotheses, even if they actually are not—as when individuals selectively attend to "confirmatory" evidence and overlook disconfirmatory evidence. Belief in illusory correlations may also hinder the detection of valid relationships.

The research on confirmatory and disconfirmatory strategies and the impact of prior belief is open to criticism. Some of these studies may have created demand characteristics that unfairly influenced results. For example, the Chapmans encouraged individuals to report relationships between variables, even when no such relationships existed. Snyder's experimental conditions may have discouraged the use of disconfirmation strategies, even given his claims to the contrary. Further, the Chapmans may have concluded prematurely that illusory correlations are a major obstacle to the detection of valid relationships. Individuals were exposed to a relatively small number of valid pairings between Rorschach responses and symptoms, and had they been allowed to examine a greater number of cases their ability to detect valid relationships may have improved. Given these possible flaws, some caution should be exercised in interpreting results; but by no means should the research be discounted or dismissed as irrelevant to science. There is relatively strong evidence that the scientist, like the layperson, often relies on confirmatory strategies (see chap. 4). Potential consequences of this bad judgment habit are too important to overlook, regardless of questions created by possible methodological inadequacies in current research. Research with the layperson suggests that confirmatory strategies may lead to judgment errors that cannot be reconciled with prescriptions for these strategies (unless one takes a decidedly nontraditional view of science). This research, though not definitive, provides suggestive leads that warrant further investigation.

CATEGORY II: LIMITS IN HUMAN JUDGMENT

Many judgment failings do not result from our having done the wrong thing, such as inappropriately using confirmatory strategies, but from our incapacity to do the right thing. An expanding body of research shows that there are major restrictions in the efficacy of our judgment, restrictions that are rooted in human cognitive limitations. A review of this research leads to one simple conclusion—individuals often perform so poorly in judgment studies because human judgment is remarkably limited. Even when functioning optimally, we are far less capable of managing complex decision tasks than we have assumed, and we are often unable even to approach our performance ideals. These conclusions probably also apply to a wide range of scientific judgment tasks and, if they do, challenge core assumptions about the scientific enterprise.

The evidence for human judgment limitations comes from two basic sources: studies exploring the relationship between the amount of information available to the judge and judgment accuracy, and those exploring performance on multiple-cue (i.e., complex) judgment tasks. Research in the latter area includes attempts to model human judges and to uncover successful implementation of complex judgment strategies. The results of this work will be discussed in detail in this section, but it is worthwhile first to summarize the unsettling picture that emerges. When performing complex judgment tasks, individuals seem incapable of properly weighting more than a few pieces of information. Pieces can be added together in a stepwise or linear manner, and sometimes simple interactions or configural relationships can be recognized. But attempts to process information at a higher level, that is, to consider more information or to analyze more complex interrelationships, are almost never successful.

Relationship between Amount of Information and Judgment Accuracy

One source of evidence for limitations in human judgment comes from studies examining the performance of judges when they are provided with varying amounts of information. Oskamp's study (1965), which was discussed in chapter 2, addressed this issue. Psychologists and nonprofessionals studied a detailed written description of an individual and then tried to predict this individual's behavior. The history was provided in four steps, at each of which about 25 percent of the information was presented. After each step, the participants responded to the set of items from which they predicted behavior. By comparing performances after each step, Oskamp could assess the relationship between the amount of information available to the judges and their predictive accuracy. As the amount of available information increased step by step, there was no associated increase in the percentage of correct predictions—additional information did not increase judgment accuracy. Nevertheless, individuals' confidence in their judgments generally increased as more information became available to them.

Other studies on this topic have yielded similar results (Golden 1964; Kostlan 1954; Sines 1959; Winch & More 1956). In a classic study, Sines found that even when very limited information was available, additional information generally improved the judgment accuracy of psychodiagnosticians only slightly, certainly far less than was typically assumed. In many cases, demographic information alone—the individual's gender, age, and marital status—resulted in judgment accuracy equal to that achieved when detailed information about history and responses to psychodiagnostic tests was made available. It was not that the additional information contained nothing of predictive use, but rather that the diagnosticians were unable to use this additional information properly (Sawyer 1966). The same result was obtained in a study by Slovic and Corrigan of horse race handicappers.[1] Beyond about five pieces of information, additional information did not increase the handicappers' predictive accuracy, although it did fallaciously increase their confidence.

There have even been instances uncovered, such as in Sines's study, in which additional information *decreased* judgment accuracy. In the comparisons analyzed by Sawyer, the addition of clinically collected data to mechanically collected data never increased judgment accuracy when data was combined clinically, but in some cases it did decrease accuracy. Apparently, newly available information was averaged in with or granted ascendancy over previously available information even when it had less predictive power. Taken together, these studies demonstrate that when even very limited information is available, additional information often does little to increase clinical judgment accuracy; at worst it decreases judgment accuracy. No situation has yet been uncovered in which judges, once given access to a limited amount of valid information, have been able to use additional information to improve predictive accuracy markedly. Nevertheless, there is a tendency to become more confident in one's judgments when additional information is made available. The failure to use additional information to increase judgment accuracy is not based on some inherent limitation in the information. Rather, it is based on an inability to weight, organize, and integrate more than small amounts of information effectively and to select from a pool the most useful information.

Performance on Multiple-Cue Tasks

Other evidence for human cognitive limitations comes from studies either comparing judges and models or examining human performance on multiple-cue tasks—tasks in which information pertaining to two or more dimensions is presented. Investigators use these tasks to study the human capacity to manage more complicated information, especially the ability to detect and utilize interrelationships among cues. This configural analysis, or consideration of patterns within the data, is basic to many forms of complex information processing, and the clinical judge purportedly has unique cognitive capacities based on the ability to perform complex configural analysis. For these reasons, performance evidenced on multiple-cue tasks provides data that are central to debates about human cognitive abilities.

Multiple-cue studies have uncovered two related findings: First, even when individuals claim to have used complex configural analysis to reach judgments, one can construct simple linear models—models that do not consider configural relationships—that adequately duplicate their performances. Second, although individuals have the capacity to utilize *simple* configural cues, this capacity does little, if anything, to improve judgment performance. No evidence has yet been uncovered to support the claim that individuals can effectively utilize complex configural cues.

Simple Models for Duplicating Purportedly Complex Judgments

The effort to construct models that duplicate the performances of human judges potentially furthers our understanding of human cognitive processes and provides information that is useful in developing procedures that improve judgment, be it clinical or actuarial. Lewis Goldberg and his colleagues at the Oregon Research Institute have been at the leading edge of the modeling research. In a landmark article, Goldberg (1968b) described possible strategies for constructing models of judges and the results of efforts to apply certain of these strategies. The central goal in modeling is to construct procedures that reproduce judges' decisions as closely as possible. From among the strategies open to Goldberg and his colleagues, they chose to start with extremely simple models and to introduce complexity as necessary. A series of studies demonstrated, however, that adding complexity was unnecessary: simple models alone could adequately reproduce judges' performances.

An example is provided by Hoffman, Slovic, and Rorer's (1968) previously described study, in which radiologists were asked to differentiate between benign and malignant ulcers. An attempt was made to model the radiologists' judgments. The selected method assessed the extent to which nonlinear terms had to be added to linear terms to construct models that duplicated the responses of judges (or accounted for the variance in their responses). Linear terms combine cues additively (a + b + c), and nonlinear terms combine cues in a configural or interactive manner. For each case, the radiologists were provided with the seven cues deemed relevant to the diagnosis. Hoffman, Slovic, and Rorer were able to construct separate models of each radiologist that adequately reproduced their judgments. The researchers then analyzed the extent to which the capacity of the models to duplicate the judges was dependent on the simple addition of cues versus configural analysis. The use of configural cues was assessed by examining all possible two-, three-, four-, five-, and six-way interactions between cues (fifty-seven interactions in all). Each model of each judge was analyzed separately. Generally, linear (i.e., noninteractive) terms accounted for most of the variance in the judges' responses. Among the fifty-seven possible interactions, the largest appearing in any of the judges' models accounted for only 3 percent of that judge's response variance. The use of simple additive cues alone usually accounted for ten to forty times as much of the response variance as did the use of even the largest interaction among cues.

Rorer et al. (1967) studied another judgment task, the decision whether or not to grant temporary liberty to psychiatric patients. The judges, selected from the professional staff of a psychiatric hospital, included six physicians, twelve nurses, three clinical psychologists, and three psychiatric social workers. Judges were provided with the six cues assumed to be most relevant in making this decision. Each cue consisted of a yes or no answer to a specific question (e.g., does the patient have a drinking problem), providing two possible responses or levels for each cue. The two levels and six cues could be combined in sixty-four different ways. By analyzing each possible combination, the researchers could determine the extent to which cues were simply added together or were combined configuratively. If configural analyses were taking place, then decisions would represent something more than the addition of the six cues and, therefore, would show systematic variations based on the interrelationships among cues. The judges evaluated 128 presumedly real, but actually hypothetical, cases that were presented randomly and included each of the sixty-four possible cue combinations, which were repeated a second time. For each case, the judges decided whether the patient should be allowed to leave the hospital grounds for eight hours.

As in the Hoffman, Slovic, and Rorer (1968) study, models of each judge were constructed that adequately reproduced the respective judge's performance. Each model was then analyzed separately to determine the extent to which nonlinear terms were necessary to account for that judge's response variance. All possible interactions were examined. The results were nearly identical to those obtained in the earlier study, with linear or noninteractive relationships accounting for most of the response variance and configural relationship accounting for little. Among all the models, the strongest configural relationship accounted for about 6 percent of the variance; for many of the models, configural relationships accounted for 1 percent or less of the variance.

Finally, Wiggins and Hoffman (1968) analyzed clinicians' use of the MMPI to differentiate neurosis and psychosis. This task, like the previous two, was not intended to make configural analysis difficult for the judges but to facilitate its use. The judges themselves claimed that they had the ability to analyze the configural cues the MMPI provided and that their interpretations were heavily dependent on such cues. Nevertheless, Wiggins and Hoffman found that the great majority of judgment variance could be accounted for by simple linear models. Even for the most configural model, the addition of nonlinear terms, which took into account configural relationships, increased the correlation obtained between the linear model decisions and the judge's actual decisions by only .03. As Wiggins and Hoffman (1968) remarked, "The judgments of even the most seemingly configural clinicians can often be estimated with good precision by a linear model" (pp. 76-77). The findings of these three studies are in close accord with others that examine the effectiveness with which linear or nonconfigural models duplicate the performance of judges (Huber, Sahney & Ford 1969; Kort 1968; Slovic & Lichtenstein 1968b; Summers & Stewart 1968; Yntema & Torgerson 1961).

What conclusions can we draw from the demonstrated capacity of simple models to reproduce purportedly complex judgments? One should not conclude that this capacity provides evidence that individuals are completely incapable of performing configural analysis. There are no assurances, nor even strong reasons to suspect, that the actual judgment strategies or steps used by individuals closely resemble those used by their models. The models recreate the judges' functions (decisions) but not necessarily their cognitive processes. For example, because of certain statistical properties, linear models can sometimes duplicate judgments that individuals actually perform using nonlinear or configural strategies. There is also strong evidence (Slovic & Lichtenstein 1971) that in at least some of the modeling studies some of the judges used configural strategies, but that their responses could still be reproduced using linear or nonconfigural models (see Birnbaum [1973] and Dawes & Corrigan [1974] for a review of the rather technical issues involved in linear modeling). There is also substantial evidence that individuals can use nonlinear strategies to weight cues, at least when considering simple problems. The general conclusion that *can* be drawn from these studies is that linear models can reproduce judges' decisions, even those purportedly reached using complex processes. As Goldberg (1968b) states, "Judges can process information in a configural fashion, but . . . the general linear model is powerful enough to reproduce most of these judgments with very small error . . ." (p. 491).

Goldberg's conclusion has at least two major implications. First, regardless of any configural strategies judges may use, most of their decisions can still be reproduced using linear models alone. Thus, there is doubt that the configural analyses performed by judges add much, if anything, to judgment accuracy, a suspicion that raises questions in turn about the sophistication of human configural analysis. If configural analysis fails to elevate performance beyond the level obtained through linear analysis alone, it is likely that the capacity to perform configural analysis is limited. Nonlinear analysis is doubtless superior for solving certain problems, such as those in which the interrelationship between three variables is critical; but human judges apparently often cannot perform this analysis.

Second, although linear models can reproduce *some* nonlinear judgments, they cannot reproduce *all* nonlinear judgments, especially more complicated ones. Nevertheless, in all studies conducted thus far, linear models have been able to account for most of the response variance displayed by human judges. If judges were performing complex configural analyses that made an essential contribution to judgment accuracy, it is very unlikely that linear models could have closely reproduced judges' decisions in every one of these studies. That linear models have uniformly duplicated the majority of judges' decisions, even when studies were designed to create conditions favorable to the clinical judge, lends plausibility to the hypothesis that individuals cannot perform complex configural analysis.

Performance on Tasks Requiring Configural Analysis

Modeling studies have provided indirect evidence about human cognitive limitations. More direct evidence is provided by the investigation of human performance on tasks requiring configural analysis or multiple-cue utilization. The results of such studies offer a third line of evidence for human cognitive limitations. As mentioned previously, there is strong evidence that individuals can perform configural analysis. For example, Slovic and MacPhillamy (1974) presented subjects with two statements about a pair of students. Based on these two statements, subjects were required to guess each student's grade point average. One of the statements described scores on a dimension common to both students: student A scored high on a test measuring achievement motivation, and student B scored low on the same test. The other statement described a unique dimension: student A scored low on a test of English skills, and student B scored high on a test of mathematical ability. In reaching judgments, subjects weighted statements describing the common dimension more heavily than statements describing the unique dimension. Statements were weighted differently, then, depending on whether they described common or unique dimensions, a form of interactive judgment indicative of a nonlinear reasoning strategy. Slovic (1966) found that the weight assigned to cues interacted with the perceived consistency of these cues. When subjects believed that the two most salient cues were consistent with each other, they weighted both cues heavily. But if subjects believed that the two cues were inconsistent, they relied on one cue more heavily than the other and exhibited greater reliance on additional cues. This interaction between assigned weights and cue consistency provided evidence that individuals were using non-linear judgment strategies. Numerous additional studies have uncovered apparent instances in which individuals use nonlinear strategies to process cues (Anderson & Jacobson 1965; Dudycha & Naylor 1966; Einhorn 1971; Gollob 1968; Lampel & Anderson 1968; Sidowski & Anderson 1967; Tversky 1969).

Demonstrations of nonlinear cue processing, however, offer little solace for those wishing to maintain that individuals can perform complex judgments competently. Although individuals can analyze configural cues, there is currently no evidence that this significantly aids judgment performance. When configural analysis is attempted, performance does not improve markedly and may even decline. When configural strategies are used, individuals also show very limited ability to integrate information, even when presented with only two or three simple cues. The configural analyses that can be performed, then, often still lead to judgment errors; they do not appear to indicate complex judgment abilities.

For example, in the Slovic and MacPhillamy study (1974) the differential weighting of statements describing common and unique dimensions was apparently directly attributable to the information processing demands of the task. Comparison along common dimension is a straightforward cognitive task, but comparison along a unique dimension requires more complex cognitive operations, such as determining the relevance of each unique dimension and performing other intermediary transformations of the information. Slovic and MacPhillamy hypothesized that individuals had difficulty carrying out the cognitive operations required to consider the unique dimension and, therefore, relied heavily on the common dimension. Their hypothesis is supported by additional reports (Payne & Braunstein 1971; Slovic & Lichtenstein 1968a; Tversky 1969) of the tendency to underutilize data requiring more complex cognitive processing when it appears together with data requiring less complex processing. The cognitive operations performed by individuals in the Slovic and MacPhillamy study may not have been linear, but neither were they complex nor a means to the optimal use of the information.

As the Slovic and MacPhillamy study suggests, nonlinear strategies may be used when the demands for processing information *increase*. The nonlinear strategies that individuals typically use may require less complex cognitive operations than do linear strategies. Adding cues, or linear cue utilization, is not always the simplest cognitive process, and certain forms of nonlinear cue utilization may simplify the judge's task. Possible examples are provided by Dawes (1964), who described disjunctive and conjunctive strategies. Although both are nonlinear, they are often simpler to perform than linear strategies. The disjunctive strategy requires that one simply determine whether the cue with the highest value meets a required level. To use Slovic and Lichtenstein's (1971) example, a scout for a professional sports team may evaluate a prospect purely in terms of his or her best attribute, be it speed, strength, or agility. A conjunctive strategy requires that one determine whether each cue meets a minimal standard. For example, a casting director for a Broadway play may require that a candidate's ability to sing, dance, and act all meet a certain standard.

The evidence for the use of such simplifying nonlinear strategies is more than anecdotal. For example, Einhorn (1971) found that faculty evaluations of graduate school applicants and students rankings of job desirability were often most accurately reproduced by conjunctive, rather than linear models. For these tasks, conjunctive strategies entailed less complex cognitive operations than did linear strategies. Einhorn went on to make a critical point, that is, that cognitive complexity and mathematical complexity do not necessarily parallel each other and should not be confused. Nonlinear strategies may be *cognitively* simpler than linear strategies, even though the mathematical operations needed to model or execute the former may be more complex than those needed to model the latter. Therefore, demonstrations of nonlinear judgment abilities by no means necessarily provide evidence that individuals can perform complex judgments. If anything, the evidence seems to support the opposite conclusion. As Slovic and Lichtenstein (1971) comment:

We find that judges have a very difficult time weighting and combining information, be it probabilistic or deterministic in nature. To reduce cognitive strain, they resort to simplified decision strategies, many of which lead them to ignore or misuse relevant information. (p. 724)

The nonlinear analyses that individuals typically use seem to represent one such type of simplifying strategy and, therefore, to reflect cognitive limitations rather than complex judgment capacities.

The finding that judges fail to improve performance using configural strategies and the evidence that typical configural judgments reflect simplifying strategies are not the only sources of discouragement for those wishing to maintain faith in complex human judgment abilities. The evidence that is currently available suggests that our ability to detect and adequately comprehend configural relationships is limited. For example, individuals often learn to recognize or detect configural relationships more slowly and less effectively than they do linear relationships (Brehmer 1969; Hammond & Summers 1965; Summers 1967; Summers, Summers & Karkau 1969), especially if they are not properly forewarned that relationships might be nonlinear (Earle 1970; Hammond & Summers 1965; Summers & Hammond 1966). Further, most demonstrations of configural analysis have occurred under contrived laboratory conditions, in which very simple tasks have been used and special supports provided for the discovery of configural relationships. This raises doubts about whether such abilities generalize to the world outside the laboratory. As Wiggins (1973) states, "Although it is possible to demonstrate nonlinear cue utilization in contrived situations, it is doubtful whether models of judgments of artificial data may be generalized to models of judgments of clinical data" (p. 179). Finally, and most important, although dozens of studies have now been conducted and ample opportunity provided, no researcher has uncovered even one instance in which any judge has displayed the ability to perform *complex* configural judgments. Even given extensive efforts at verification, no judge's claim to complex configural judgment abilities has been supported; yet the obverse is clearly not true.

Conclusions

The investigation of models that duplicate judges' decisions and direct study of configural analysis have not produced evidence for complex judgment abilities. Combined with studies demonstrating our limited capacity to integrate and utilize more than a small amount of information, this body of research strongly suggests that individuals are far less capable of performing complex judgment tasks than previously assumed. Perhaps some judges in some studies performed complex configural analysis, but these instances were not detected. Even if it occurred, any gains realized from these strategies appear to have been minimal. If significant gains had been achieved, the configural clinician would almost certainly have surpassed the actuarial method in at least some of the numerous comparative studies, given that many of these studies used actuarial methods that simply added cues in a linear fashion. This result has never been obtained.

We can thus conclude that, if individuals can perform complex configural analysis (and there is no evidence to support this claim), they can do so in an unreliable or minimally valid fashion—one that provides no significant gains in performance beyond the level obtained using linear strategies. Further, the capacity to perform configural analysis, or any other judgment abilities the clinician *might* possess, does not appear to provide a means for surpassing actuarial judgment methods. Although some investigators claim that the clinical judge might be able to use configural analysis to outperform the actuary in certain situations, there are currently no data to support this claim. Goldberg's (1968b) earlier statement remains equally valid today:

While Meehl (1959) has suggested that one potential superiority of the clinician over the actuary lies in the human's ability to process cues in a configural fashion, it is important to realize that this is neither an inherent advantage of the human judge (i.e., the actuary can include nonlinear terms in his equations), nor is this attribute —in any case—likely to be the clinician's "ace in the hole." If the clinician does have a long suit—and the numerous clinical versus statistical studies have not yet demonstrated that he has—it is extremely unlikely that it will stem from his alleged ability to process information in a complex configural manner. (p. 491)[2]

The point then, relative to the general issues raised about human judgment capacity, is twofold. First, there is considerable evidence that individuals are far less capable of performing complex judgment tasks than previously assumed. This is probably the result of limited cognitive capacity. Second, even if complex configural analysis is possible, clinical judges can be outperformed by the actuary. Given the current balance of the evidence, any individual making claims to the contrary must bear the burden of proof.

REFLECTIONS ON LACK OF INSIGHT

Assuming that it is true that human judgment abilities are extremely limited, at least relative to common beliefs about these abilities, one might ask why we have consistently overestimated our capacities. Subjectively, we experience ourselves as competently performing complex judgments, or believe we have witnessed others doing so. But subjective impressions, whether we are attempting to portray our internal cognitive operations or some aspect of external reality, can be misleading; by themselves, they obviously do not represent definitive scientific evidence.

Actually, we flatter ourselves to think we can apply conscious reflection to gain full insight into our reasoning operations. Any "apparatus" that explains or fully understands something must be more complex than the thing explained (Hayek 1955). Therefore, to fully understand our own reasoning, our mind would have to be more complex than itself, an obvious absurdity. Aside from abstract arguments about the limits of self-reflection, voluminous experimental evidence has been gathered since Wundt first conducted laboratory studies of subjective perceptual experience in the late 1800s, showing that introspective reports are laden with errors (for a review of Wundt's work, see Marx & Hillix [1973]). Wundt's subjects had to undergo lengthy training to *learn* to give assumedly accurate introspective reports. Diverse contemporary research also shows that subjective impressions about our judgment processes or abilities are often inaccurate. In the Oskamp (1965) study, individuals became increasingly confident in their judgments, a subjective experience discordant with their failure, in actuality, to make more accurate judgments. Slovic (1976) and Fischhoff, Slovic, and Lichtenstein (1977) provide instances in which individuals were far more certain about their judgments than was justified. For example, the conclusions of subjects were often wrong, even when they were very certain they were right. Various researchers (Hoepfl & Huber 1970; Hoffman 1960; Oskamp 1962; Slovic 1969) have compared subjective reports and objective measures of the weights individuals assigned to cues when interpreting data. Individuals consistently overestimated the importance they placed on minor cues (i.e., subjective reports of weights exceeding objective measures of weights) and consistently underestimated their reliance on a few major cues.

Additional evidence for the unreliability of introspective reports that is perhaps the most relevant to the current discussion comes from the work of Nisbett and Wilson (1977a, 1977b; Wilson & Nisbett 1978). These researchers obtained a measure of access to internal operations by examining individuals' ability to recognize the factors that influenced their judgments. The Nisbett and Wilson studies were designed to minimize the forms of defensiveness that might lead individuals to disguise the causes of their behavior or judgments from the experimenters or themselves. For example, attempts were made to ensure that accurate and inaccurate responses were of equal social desirability. Nevertheless, individuals were remarkably inaccurate when reporting the factors influencing their judgments. They reported that factors that had been independently demonstrated as having a marked effect on their judgment had not at all influenced their thinking. Factors independently demonstrated as having no significant effect on judgment were reported to have been influential. Nisbett and Wilson's work placed in doubt all studies in which individuals' subjective reports of their problem-solving strategies were assumed to reflect accurately the internal operations they actually used.

We have no reason to conclude, then, that our subjective impressions that we or others can perform complex judgments or configural analysis establish that individuals can really do these things. This is not to question whether subjective *experiences* exist or can potentially be reported accurately, but whether these experiences or impressions accurately represent the cognitive operations we perform. Although disconfirmatory evidence for valid introspections obviously does not prove that we can never accurately identify our internal cognitive operations, it does argue against uncritical acceptance of subjective impressions. And, certainly, subjective impressions alone are by no means adequate to discount counterintuitive claims about limits in our judgment abilities. Therefore, colleagues who insist that the judgment literature is irrelevant because *they know* they (or others) can perform complex judgments fail to realize that they may not know of what they speak. The question of whether we can or cannot perform complex problem solving cannot be resolved by reference to subjective experience; rather, it requires the study of human performance on judgment tasks.

The evidence for mistaken subjective impressions of internal processes lends plausibility to the argument that we lack insight into our own judgment limitations, but it does not explain the origin of our false self-assessments. Perhaps these false beliefs originate through subjective experience and are maintained by the use of confirmation strategies when evaluating our judgment abilities. When individuals form a hypothesis, subsequent evidence, even if it is ambiguous, is interpreted as supportive of one's hypothesis. Selective attention to and overvaluation of confirming instances leads to the maintenance of these false beliefs (Anderson 1972; Chapman & Chapman 1967, 1969; Snyder 1978).

In the realm of complex decision making, we are rarely provided with clear evidence or feedback about our ability or inability to perform complex judgments. Most commonly, we find that our judgments are correct at times and incorrect at other times; this is clearly ambiguous evidence. As with the individuals who evaluated the validity of their prior hypotheses in the Chapman and Chapman (1967, 1969) studies, however, selective attention to correct judgments and the greater weight accorded to confirming instances would lead one to believe that support had been found for one's capacity to adequately perform complex judgments. If this process occurs when we analyze our abilities to reason complexly, it could result in the maintenance of false beliefs. What remains unexplained is the initial formation of subjective impressions that we can adequately perform complex judgments. I leave this question to the sociologists, psychologists, and others who have specialized in the study of our unrealistic self-appraisals.

CHAPTER 4

Human Judgment and Science

Any attempt to foretell the extent to which the judgment findings generalize to scientists is as intriguing as it is hazardous. An extreme position will be put forth here, in part as a challenge, but mainly because I believe it to be more true than not. I hypothesize that for any judgment or cognitive limitation demonstrated to be common to an intelligent adult population, parallel instances will be found with scientists across fields.[1] Fallacious or limited reasoning by scientists will of course occur in the context of a different subject matter — the body of knowledge relevant to the scientist's work. Therefore, the exact manifestations of reasoning errors and cognitive limitations will vary, but their presence will not change. For example, the lay person often ignores base rates and consequently makes judgment errors, as in disregarding a national survey demonstrating a car's excellent maintenance record if a neighbor reports that his or her car of this brand was unreliable. In a parallel fashion, the scientist may ignore base rates and consequently make judgment errors, as in using diagnostic signs that decrease hit rate relative to that which would be achieved using base rates alone. As the judgment errors and cognitive limitations of scientists are exposed, I further expect that a substantially revised picture or description of science will emerge, one that will have major prescriptive implications.

REASONS TO HYPOTHESIZE GENERALIZATION
TO SCIENTISTS

There are at least three reasons to hypothesize that judgment findings will generalize to scientists. These include evidence for continuity between scientists and nonscientists, similarities between the tasks used in judgment studies and tasks performed by scientists, and the results thus far obtained in judgment studies with scientists and quasi scientists.

Continuity between Scientists and Nonscientists

There may be far greater continuity between scientists and nonscientists than is commonly assumed. First, the lay person or nonscientist, at times, seems capable of functioning much like the stereotypic scientist. Nonscientists often demonstrate the capacity to implement scientific problem-solving strategies that are generally thought to be only used by the scientist. Perhaps the most compelling examples come from the work of Piaget (1952). Piaget has clearly demonstrated that individuals commonly solve problems using principles of logic and procedures classically associated with the scientific method. Individuals who have reached the highest state of Piagetian reasoning, formal operations, are capable of using such strategies. This probably includes 50 percent or more of the adult population, at least in the United States (Kuhn et al. 1977). When solving everyday problems, formal reasoners often construct hypothetical instances as an aid to the formulation and assessment of hypotheses. "If . . . then" reasoning is common, and a hypothetico-deductive approach to the testing of hypotheses is competently applied. Real-world tests are conducted, in which efforts are made to control conditions and to systematically isolate and manipulate variables.

Piaget discovered these abilities when studying subjects' performances on a series of tasks. On one of these, individuals were provided with metal bars that varied along a number of dimensions, such as length, thickness, and type of material. Subjects were then asked to determine how each dimension affected the pliability of the bars. To solve this problem, they had to isolate and vary one dimension at a time, while holding all other variables constant. They then had to proceed systematically to test all of the variables in this manner. Formal reasoners have no difficulty on this task, easily obtaining the correct solution through the application of scientific problem-solving strategies.

Although formal reasoners show well-developed scientific problem-solving skills, these skills apparently provide little or no protection against the judgment errors and limitations that have been documented. Unfortunately, researchers have not directly analyzed the relationship between stages of Piagetian reasoning and performance on judgment tasks. Given the high proportion of formal reasoners in the adult population, however, and the consistency with which judgment errors have been demonstrated in many of the subject groups studied, many of the individuals displaying erroneous judgment were probably formal reasoners. In fact, the literature contains not a single report of a large subgroup of subjects avoiding judgment errors.[2] If formal reasoners avoided errors, one would expect many such reports. In some instances, perhaps some formal reasoners did not make the judgment errors committed by those at lower stages of reasoning, with this effect being averaged out when the performances of the entire group were combined. But the obvious inference to be drawn from the currently available evidence is that formal reasoners are prone to most or all of the major judgment errors and limitations that have been documented. Although many nonscientists can apply scientific problem-solving strategies, such abilities appear insufficient to prevent many judgment errors. And these strategies offer no safeguards against judgment failings rooted in cognitive limitations.

If the nonscientist's use of formal reasoning strategies does not prevent major reasoning failures, what are the grounds for assuming that the scientist's use of similar strategies accomplishes this aim? Although the strict application of experimental procedures may offer the scientist certain advantages over the nonscientist, the processes involved in obtaining scientific knowledge require reasoning and not the simple tallying of research results. Detecting or discovering relationships in complex sets of data and evaluating scientific constructions are, in significant part, exercises in reasoning, exercises for which few conventional safeguards exist. Under conditions in which the scientist must apply reasoning powers to infer meaning from research results, he or she, like the lay person, has available the reasoning strategies associated with formal operations. Such capacities do not seem adequate to prevent judgment errors, especially those associated with cognitive limitations in processing information. But there is no higher stage of reasoning to which one can appeal. As far as we know, individuals do not attain higher stages; the scientist and nonscientist alike are bound by the limits of formal reasoning.

Overestimations of the scientist's intellectual superiority may be a second factor underlying the unrecognized level of continuity between the scientist and nonscientist. Genius, or at least unusual mental power, is often believed to be common among scientists. We have reason to question this stereotype, which may have emerged and been perpetuated by the extraordinary attention directed towards the rare scientific geniuses that have enhanced their respective fields. Many stereotypes are apparently formed and maintained through this selective attention to a few highly evident cases. An example is the belief that adolescents generally experience extreme life turmoil. Investigations have not supported this stereotype, which is apparently based on a few well-publicized cases of rebels without causes (Weiner 1970). There is little doubt that scientific genius exists or that most scientists have above average reasoning powers, but it is quite unlikely that most scientists are geniuses.

These doubts are supported by the investigations of Roe (1953), Bean (1979), and Mahoney and Kimper (1976). As part of a study of eminent physicists, Roe conducted intellectual assessment. Many of the physicists obtained IQ scores that did not approach the genius level. The studies of Bean and of Mahoney and Kimper found that many scientists do not adequately comprehend certain of the basic principles of logic underlying scientific method. But questions about the scientist's superiority extend beyond the issue of sheer intellectual power, including also the selection and use of reasoning strategies. Available evidence suggests that scientists and nonscientists similarly use reasoning strategies that violate what are often considered scientific ideals.[3] For example, Mahoney and DeMonbreun (1977) found that scientists predominantly used confirmation strategies on a problem-solving task. Mahoney and DeMonbreun's demonstration may be the proverbial tip of the iceberg.

In summary, we probably have underestimated the nonscientist's capacity to use scientific problem-solving approaches and overestimated the scientist's superiority in intellectual strength and selection of reasoning strategies. Far greater continuity may exist between the two groups than is generally recognized. The greater the continuity, the more likely that scientists commit the same reasoning errors and are bound by the same judgment limitations as lay people.

Task Similarity

A colleague has expressed doubt that judgment findings will generalize to scientists because of the limited similarity between tasks used in judgment studies and those performed by scientists. This argument is perplexing. Analysis of the tasks used in the judgment literature, if anything, provides a second grounds for hypothesizing generalization to scientists. Many of these tasks have strong positive analogy to the tasks commonly performed by scientists. For example, extensive study has been conducted of categorization or assignment to a diagnostic class (e.g., Hislop & Brooks 1968; Levine 1969). Categorizing or classifying, a common scientific activity, occurs both formally in the construction and utilization of conventional taxonomies and informally in the conceptualization or interpretation of phenomena. The maintenance or modification of hypotheses following exposure to new data has also frequently been studied (e.g., Chapman & Chapman 1967, 1969; Ross, Lepper & Hubbard 1975). Surely this has something in common with hypothesis testing in science. Other studies have been directed towards causal reasoning (e.g., McArthur 1972; Storms 1973; Taylor & Fiske 1975), which infuses nearly all scientific investigations.

In these and other available examples, clear parallels exist between the tasks used in the judgment studies and the tasks performed by scientists. Therefore, there is little basis to assume, out of hand, that judgment findings will not generalize because of a lack of task similarity. I do agree, however, that a major dissimilarity exists between tasks used in judgment studies and tasks performed by scientists. The latter are frequently far more difficult and complex. One example, which will be further elaborated below, is theory evaluation. If the tasks required of scientists hold parallels to those used in judgment studies, but are far more difficult and complex, this point of dissimilarity leads to the opposite conclusion of that reached by my colleague. If anything, the differences between commonly studied judgment tasks and scientific tasks should increase the likelihood that judgment errors and limitations will be found in science.

Initial Findings from Studies of Scientists

In addition to inferential grounds, direct evidence supports the hypothesis of generalization to scientists. A growing list of investigations and observations has documented the judgment errors and cognitive limitations of scientists. This documentation comes from the study of scientists' performances on analogue or quasi-scientific tasks and more importantly from the study of everyday scientific practices. Many studies and observations have exposed bad judgment habits and the failure to use normative guidelines. Meehl and Rosen (1955) observed that psychologists, including those with rigorous methodological training, often draw erroneous conclusions from psychological tests because they overlook or underutilize base rate information. Tversky and Kahneman (1971) found that skilled methodologists make judgment errors associated with belief in the law of small numbers and that many experiments are wasted and research findings misinterpreted because of failure to appreciate the characteristics of small samples. Chapman and Chapman (1967, 1969) discovered that prior hypotheses and use of confirmatory strategies by psychologists leads to belief in illusory correlations. In addition, a plethora of sociological studies (e.g., Mitroff 1974) describes violations of what many consider to be normative procedures.

Poor judgment performance and cognitive limitations have been demonstrated in numerous studies with scientists and quasi scientists (i.e., individuals with training in basic scientific methodology and extensive experience in an applied scientific field, such as medicine). Many of the scientists that Bean (1979) studied did not comprehend certain basic rules of scientific logic. Mahoney and Kimper (1976) found that many scientists do not recognize disconfirmation as a valid form of reasoning. Studies of medical diagnosis (e.g., Hoffman et al. 1968) and psychiatric diagnosis (e.g., Goldberg 1965, 1968b) provide additional examples of poor performances and unsuccessful attempts to execute complex judgment strategies.

Research with scientists and quasi scientists, then, has convincingly demonstrated the presence of bad judgment habits, failures to use normative rules as we best understand them, and limitations in the ability to perform complex judgments. There is little doubt that scientists commit judgment errors and have cognitive limitations. The extent and possible manifestations and consequences of judgment errors and cognitive limitations are at issue, but not their existence. The research to date suggests that such errors and limitations are extensive and that their consequences are substantial. Therefore, the aim of the following discussion is to anticipate the extent and consequences of judgment errors and limitations, not to establish their existence.

JUDGMENT ERRORS STEMMING FROM BAD HABITS AND FAILURE TO USE NORMATIVE METHODS

Instances and Implications of the Base Rate Problem

Might failure to consider base rates be a source of judgment error in scientific decision making? This appears likely in the applied sciences, especially in the performance of diagnostic or classification tasks. Signs associated with specific diagnoses or taxonomic categories are frequently used as aids or determinants for classification. In many situations, use of the sign—given its reliability for correctly identifying an instance of the relevant event or category and the frequency with which instances of the event or category occur—actually decreases the hit rate (the frequency of accurate identifications) relative to that which would be achieved using base rates alone. In these situations, if we want to maximize hit rate, the sign obviously should not be used. Surely there are cases in the applied sciences in which base rates are insufficiently weighted and, as a result, signs are used that actually decrease hit rates.

Reliance on the electroencephalogram (EEG), at least as sometimes used to identify children with minimal brain dysfunction, may be one example. Normal children often obtain abnormal EEGs (Capute, Niedermeyer & Richardson 1968; Petersén & Eeg-Olofsson 1971). No reliable method has been developed to differentiate the abnormal EEG records of normal children and the abnormal EEG records of children with minimal brain dysfunction (Hughes 1976). The frequency of abnormal EEG findings in a population of normal children varies widely in relation to several factors. Some researchers have reported rates as high as 70 percent.[4] For this example, I will assume conservatively that 10 percent of normal children obtain abnormal EEGs. It has been estimated that minimal brain dysfunction exists in 5 percent of children and that about half of these children obtain abnormal EEGs, or those correctly indicating brain dysfunction (Wender

1971). The above frequencies, applied to a representative sample of one hundred children, are summarized in figure 2. Box A refers to

Electroencephalogram Results

Child	Normal	Abnormal
Normal	**A** $n = 85$ Correct negative	**B** $n = 10$ False positive
Brain dysfunction	**C** $n = 2$ False negative	**D** $n = 3$ Correct positive

Figure 2. Electroencephalogram results for a representative sample of one hundred children.

correct negative cases—normal children with normal EEGs; box B, false positive cases—normal children with abnormal EEGs; box C, false negative cases—abnormal children with normal EEGs; and box D, correct positive cases—abnormal children with abnormal EEGs. By examining the frequencies, one can see the application of the previously stated principle: that the use of a sign with a false positive rate exceeding the frequency of the event to be predicted results in a lower hit rate than would be obtained if predictions were founded on base rates alone. When using the EEG, the overall number of correct identifications—summing the correct negative cases in box A and the correct positive cases in box D—is eighty-eight; thus, an 88 percent hit rate is achieved. If one were to use base rates alone, one would simply classify all children as normal. All ninety-five normal children would be correctly identified and the five abnormal children incorrectly identified, therefore achieving a hit rate of 95 percent. More than twice as many children (twelve to five) would be misdiagnosed if one used the EEG. Whether it is desirable to avoid missing cases in which minimal brain dysfunction exists at the expense of misdiagnosing a greater proportion of normal children as brain damaged could be argued either way, but this is irrelevant to the point raised here. What is relevant is the mistaken assumption that the use of the EEG increases hit rate, a judgment error rooted in the failure to adequately consider base rate information.

Judgment errors associated with insufficient consideration of base rates are probably not confined to diagnosis or classification in applied settings. Consider the manner in which the unavailability or underutilization of base rate information further complicates the following two questionable, but common, practices in the social sciences. First, take the procedure that is commonly followed when conducting post hoc analysis. When originally planned statistical tests do not produce significant results, researchers often begin combining the data in new ways or applying other statistical tests, continuing until *some* significant result is produced. What are we to make of significant results obtained in this way? What method can be used for determining the probability of (or the base rates for) obtaining significant results under such conditions? Clearly, the standard .05 or .01 level of probability inaccurately represents the actual base rates for obtaining significant results by chance alone when one has tried many ways of combining the data and many significance tests. The problem is greatly confounded both by the rarity of sound replication in the social sciences, which eliminates a potential safeguard against fortuitous findings, and by the frequent failure to fully report the procedures followed in the post hoc analysis. We are left with insufficient information or means to calculate the probability that many of our supposedly established findings in the social sciences are actually artifacts of chance. The probability statements provided by statistical tests often represent extreme underestimations of the likelihood of obtaining "significant" results by chance alone.

Second, the tendency to underestimate the frequency of false positive experimental results is probably heightened both by our limited exposure to experimental results in which the null hypothesis is supported and by the unavailability of information about the frequency with which such experimental results are obtained. The strong trend in U.S. journals to reject studies finding support for the null hypothesis further exacerbates this problem. Not only do many "nonsignificant" findings never appear in the literature, but, at an unknown rate, researchers never submit such studies for publication, either believing that this virtually ensures rejection or indicates some flaw in design. Our limited information about the frequency of studies supporting the null hypothesis makes it impossible to interpret the meaning of standard tests of significance. If this information were available, one could calculate the actual base rates at which spurious significant results were likely to occur.

For example, suppose we know that ten different experiments were conducted testing the relationship between sun spots and aberrant human behavior, all using the .05 level of significance, and that one of these was significant. We would know that the probability of obtaining one significant result among a series of ten experiments, regardless of the true state of nature, was not .05 but at least .50. But without access to this information, such probabilities cannot be calculated. Further, as Tversky and Kahneman (1973) have demonstrated, the ease with which one can recall instances of an occurrence partially determines judgment about the frequency of that occurrence. Therefore, one would hypothesize that the far greater availability of significant findings results in an overestimation of the frequency of their occurrence relative to nonsignificant findings. Suppose that a series of experiments explored a hypothesized relationship and that most of them resulted in nonsignificant findings. Most of the studies published on the topic, however, were those with significant results. The tendency would be to underestimate the percentage of nonsignificant findings and, thereby, underestimate the probability that the significant findings were a product of chance. The unavailability of base rate information pertaining to nonsignificant findings, then, would lead to underestimations of the likelihood that significant findings occurred by chance alone, resulting in many false positive errors of interpretation.

As a possible instance of judgment errors resulting from the unavailability of base rate information, one can easily imagine a situation in which many researchers are trying to uncover a significant relationship between a certain set of variables. Even if no relationship exists, spurious significant findings are likely to appear as repeated experiments are carried out, given the standard .05 level of significance. And it is this one or these few significant experimental results that are published and publicized. If base rates are ignored, one might assume that there is a 95 percent chance that the results are indeed significant. If base rates had been considered, it might be shown that the finding of significant results, regardless of external reality, was almost ensured. There is, nevertheless, no objective information upon which to derive an accurate estimation of base rates. The scientist is left to his or her individual devices in formulating an intuitive impression of base rates, certainly not the public or objective process one strives for in science. The far greater availability of concrete instances of significant findings further works against the formation of accurate impressions. The end result is that individuals apply individual judgment guidelines, few or any of which are likely to exploit maximally the valuable information potentially available through knowledge of base rates, and many of which may be seriously in error. That so little attention has been directed towards gathering base rates for the occurrence of nonsignificant results may also indicate a general failure to appreciate the crucial importance of base rate information.

Although I will give limited attention to prescriptive implications arising from bad judgment habits and normative violations, it is difficult to ignore them completely. In scientific fields that frequently use "inferential" statistics, investigators could begin to develop methods that estimate the probability of obtaining significant results when various procedures are used to perform post hoc analysis. They could also extend current information about the frequency of unpublished nonsignificant results and its bearing on the likelihood of making false positive errors when interpreting experimental results. One might study scientists' estimates, under commonly encountered conditions, of the base rates for obtaining spurious significant results and determine the extent to which this shapes their conclusions when performing such judgment tasks as evaluating the merits of specific research results. Investigation could also be directed towards identifying situations in which knowledge of base rates would lead to clear decisions: for example, circumstances in which significant results are virtually guaranteed and more rigorous tests are needed. Beyond significance testing and classification in applied settings, it would be useful to explore other situations in which important base rate information is underutilized or ignored, resulting in judgment errors.

Instances and Implications of Belief in the Law of Small Numbers

Although Tversky and Kahneman (1971) have shown that belief in the law of small numbers is common among scientists, at least in the social sciences, the consequences of this bad habit are probably less serious than those of many other bad habits; almost certainly they are less significant than those stemming from judgment limitations. Errors associated with belief in the law of small numbers are more likely to be detected by the research community and, once detected, corrected with relative ease if the appropriate decision aids are applied. Nevertheless, because these errors are common they result in many invalid or unsupported conclusions, which introduce significant noise and confusion into the literature. Further, when individuals unknowingly publish invalid conclusions, they can mislead others into adopting invalid beliefs. As the work of Chapman and Chapman shows (1967, 1969), these invalid beliefs may subsequently be maintained even in the face of disconfirmatory evidence.

According to Tversky and Kahneman (1971), belief in the law of small numbers results in a number of common misconceptions and judgment errors. Their conclusions are not based on inferences drawn from studies with lay people; they come directly from responses by scientists to questionnaires and from observations of common scientific practices. They report that researchers often hold unreasonably high expectations about replicating significant results when the sample size of the replication study is equal to or less than that of the original study and that researchers have undue confidence in early experimental results obtained from only a few subjects. As a consequence, they may unwittingly gamble research hypotheses on small samples without realizing the almost prohibitive odds against detecting the effects being studied, even if these effects exist. Further, even when researchers conduct studies in which the likelihood of discovering significant results is extremely low, due to lack of power in experimental design, they rarely attribute unexpected nonsignificant results to insufficient sample size or normal sample variability; rather they seek (and find) causal explanations for their results. Frequently, then, hypotheses are generated to explain nonexistent relationships.

Shortly after first reading Tversky and Kahneman's 1971 article, I attended a presentation by a researcher that exemplified many of the potential judgment errors associated with belief in the law of small numbers. This researcher was "testing" Feingold's (1975) hypothesis that certain food additives increase the activity level and decrease the attentional capacities of some children. Children were identified whose parents had placed them on diets free from these food additives because of assumed overactivity and attentional problems. The children's activity level and attention were assessed while they remained on their restricted diets, and then again following ingestion of liquids loaded with one of the additives to which they were supposedly sensitive. The researcher initially planned to test a large number of children. As is common in clinical research, however, practical problems substantially reduced the sample size, in this case to a total group of twelve children.

No significant differences were found between the on-diet and off-diet (loading) conditions. The researcher interpreted this result as indicating that food additives do not influence activity level or attention. He did qualify his results, putting forth a series of possible reasons why no differences were found: the children came from high socioeconomic families and could differ from the general population along any of a number of variables possibly related to additive sensitivity; or the specific additive used in the study is not among those that influence activity or attention. But this researcher's interpretation of the results and proposed explanations for the nonsignificant findings were, at best, very weakly supported by the experiment. Given the small sample size and variability among the subjects, the chances were extremely low that a significant relationship would have been detected, regardless of the true state of nature. In a textbook instantiation of Tversky and Kahneman's arguments, this researcher failed to realize that his nonsignificant results were very possibly due to low sample size and weak design power. The results, therefore, were probably noninformative rather than indicative of nonsignificance. The hypothesis was never really tested, or it was tested poorly, rather than disconfirmed. Nevertheless, disconfirmation was assumed and causal explanations invented.

Awareness of potential judgment errors associated with belief in the law of small numbers, as illustrated above, may help us to recognize instances in which corrective measures are required. For example, the belief in the law of small numbers, at least in the social sciences, apparently results in many wasted experiments—ones with such limited power in design that the detection of significant results is highly unlikely. The cost of these experiments goes beyond wasted effort. They can lead to either or both false negative conclusions—the mistaken assumption that a significant relationship does not exist, when it actually was not detected because of inadequate power in experimental design; and false positive judgment errors—the mistaken assumption that causal relationships exist that explain unexpected nonsignificant results, when they actually were an artifact of small sample size.

Instances and Implications of Confirmation "Bias"

Although the term "confirmation bias" is common in the judgment literature, I purposely enclose the latter word in quotation marks in the heading above. It is not assumed here that the use of confirmatory strategies necessarily represents a scientific-methodological sin. For example, Kuhn (1970) has presented a strong argument that closed-mindedness, a phenomenon that appears closely linked to confirmatory strategies, is advantageous at certain points in the development of a scientific school. It is assumed, however, that reasoning strategies have important consequences; for example, different strategies are probably more or less functional depending on the particular characteristics of the scientific puzzle to which they are applied and the state of a scientific school. It is further assumed that scientists do not uniformly use disconfirmatory strategies and that we know almost nothing about either the actual frequency with which disconfirmatory and confirmatory strategies are applied and the conditions under which they are applied. Finally, it is assumed that potential discrepancies among the strategies that are prescribed, believed to be used, and actually used are of relevance.

As with underutilization of base rates and belief in the law of small numbers, more than inferential grounds exist for assuming that scientists, at times, rely predominantly on confirmatory strategies. In reviewing the literature, one has the sense that scientists are well aware of this tendency, feel little remorse about freely abandoning disconfirmatory strategies, and believe that philosophical objections to such practices are based on ivory tower conceptions of science that ignore practical realities. Some accounts of reliance on confirmatory strategies come from field studies and interviews. One of the scientists in Mitroff's (1974) field study of the Apollo moon program offered the following statement after observing a colleague's refusal to acknowledge disconfirmatory data:

They just don't change, do they? But then, perhaps if I were honest with myself, I'd say I haven't changed much either and again I'm not sure that it is always bad for science for scientists not to change easily, although it can be extremely dangerous and irritating at times (p. 167)

Other demonstrations of reliance on confirmatory strategies have come from the laboratory. Mahoney and DeMonbreun (1977) presented scientists and clergy with a riddlelike task requiring the generation and testing of various solutions or hypotheses and compared their problem-solving strategies. Surprisingly, they found that the clergy used disconfirmatory strategies more often than the scientists. Both groups, however, relied primarily on confirmatory strategies. Weimer (1977) criticized certain features of this study, pointing out essential differences between a game or riddlelike task, in which the experimenters know the solution and individuals receive unambiguous feedback about right and wrong answers, and the practice of science, in which no one knows the solution and no one receives unambiguous feedback. Nevertheless, the results of the Mahoney and DeMonbreun study and the observations gathered from field studies are strikingly congruent. Taken together, this work suggests that scientists may often rely predominantly on confirmatory strategies.

Although the use of confirmatory strategies clearly violates certain philosophical decrees, is there any experimental evidence that such strategies impede, as we best understand it, the march towards scientific truth? If, as Kuhn and others argue, close-mindedness may at times be functional, why categorically execrate confirmatory strategies? In-depth exploration of the issues raised by this question is untenable here, but we can discuss some of the more provocative implications that can be drawn from judgment studies. The Chapmans' research (1967, 1969) demonstrates that prior belief can lead to marked distortions of relatively simple data and that this probably occurs because of reliance on confirmatory strategies. If confirmatory strategies do result in such distortions, one can question the compatibility of confirmatory approaches, at least under certain conditions, with the march towards scientific truth (as we best understand it). To explore this question, let us examine a scientific activity in which confirmatory strategies possibly lead to judgment errors and discuss how one might investigate this possibility.

Exponential expansion in the number of scientific articles and journals has given the review article an increasingly important function. Outside of their narrowing areas of expertise, researchers frequently depend on such articles to acquire familiarity with relevant areas of investigation. But the study of illusory correlations (Chapman & Chapman 1967, 1969; Golding & Rorer 1972; Starr & Katkin 1969) raises serious doubts about dependence on review articles. Chapman and Chapman demonstrated that individuals expecting to find relationships in data tend to do so, even if no supporting evidence actually exists. Once one forms assumptions about relationships they become highly resistant to disproof, sometimes resulting in grossly distorted interpretations of subsequent information.

Is it possible that reviews contain these distortions of information? In most cases, the reviewer forms one or a series of hypotheses well before covering all of the literature or even before beginning the formal literature search. The tendency to form presuppositions is heightened by the frequency with which reviewers have expertise in the relevant area and previously publicized commitments to specific positions. But even if a researcher can enter a field empty of previously established positions, factors extraneous to the data often strongly shape subsequent impressions. Not only the external factors identified by sociologists, but even such extraneous variables as the order in which one is exposed to the data can have a major impact (see Asch [1946] for a classic demonstration of the impact of order effects; more recent discussions and overviews are provided by Anderson 1974; Jones & Goethals 1972). As an illustration, assume that Jones starts by reading a study by Smith, who argues that sunspots are associated with aberrant human behavior. Jones becomes interested in learning more and possibly even writing a review article. He likely proceeds by obtaining the articles cited in Smith's study. These articles probably present views consistent with Smith's. Further, Smith has also likely made statements about the studies she has cited, to the effect that the articles that support her position are the "good" ones and the articles that are nonsupportive are the "bad" ones. The result is that, even if Jones did not hold a position before beginning the review, these and other factors common to the process of conducting a review begin to shape the formation of a position well before the literature survey is complete.

If these preliminary positions are formed, how might they influence the subsequent thinking of a reviewer? Assuming that the Chapman and Chapman findings can be generalized, once these opinions or hypotheses have been formed (all other things being equal), the subsequent search is far more likely to convince the reviewer that his or her initial impression has been confirmed rather than disconfirmed. At least two factors underlie the tendency to assume confirmation. The first is the complexity inherent in the process of conducting a review and the unavailability of standardized procedures, which together necessitate many subjective decisions. One must determine the boundaries between relevant and irrelevant studies, boundaries that are rarely distinct. One must also decide what weight is to be given to each study and make related judgments about the methodological adequacy of studies. For example, in virtually all reviews, certain studies are dismissed because of perceived methodological flaws. One must also invent some means for integrating findings across studies that frequently have major differences in method and definition. That multiple and complex decisions must be made for which no refined and systematized procedures are available ultimately means that the reviewer must rely on his or her own devices.[5] To the extent that scientifically established and public procedures are not available, the reviewer is more susceptible to the influence of bad judgment habits, such as overreliance on confirmatory strategies.

The second factor encouraging one to assume confirmation is the high frequency of contradictory results, at least in the social sciences, which makes many supportive articles available regardless of the position one takes. The almost uniform availability of confirming evidence greatly increases susceptibility to overreliance on confirmatory strategies and the associated distortions in judgment that Chapman and Chapman have described. An excellent example of the operation of this factor and the question it raises about the reliability of review articles is provided by research in psychotherapy outcome (for a general overview, see Garfield & Bergin [1978]). Hundreds of psychotherapy outcome studies have been conducted, many of which appear contradictory. Nevertheless, not only have different reviewers used this research to support diametrically opposed positions, but there have even been clashes between individuals who have reviewed the many review articles. Regardless of one's position, dozens of supportive articles can be found, and nonsupportive articles can be dismissed or reinterpreted as supportive.

How might one investigate the possibility that confirmatory strategies are a potential source of judgment error in the review process? One approach to this problem would require the identification of two groups of scientists holding clearly opposing positions on a given topic. Preferably in a controlled experimental setting, these two groups would then be exposed to actual or hypothetical data and asked to make certain judgments. For example, one could begin by compiling a pool of studies addressing the disputed topic. Studies meeting two criteria would be selected for inclusion in the experimental phase of the investigation. First, within and between group consensus would have to exist on the position supported, at least when the studies are taken at face value. It would probably be necessary to ask the scientists temporarily to suspend judgments about methodological adequacy when rating the studies—that is, to proceed as if all the studies were methodologically sound. This request would potentially restrain what might be an overpowering tendency to view studies as consistent with one's position and might allow one to collect enough studies meeting the first criterion. If this procedure did not succeed, one could provide access to the results of the studies only and temporarily withhold information about methodology. Second, independent judges who were unfamiliar with the issues under dispute would rate the methodological adequacy of the studies along several dimensions. Only studies that could be rated reliably would be retained. If a greater number of studies supporting one or the other position met the two criteria, studies would be randomly eliminated until an equal number supporting each position was obtained.

Once the selection and methodological rating of the studies were completed, the two opposing groups of scientists would then rate the methodological adequacy of these studies along the same dimensions evaluated by the independent judges. The ratings generated by the groups could be compared with each other and with the ratings generated by the neutral judges. Would there be a systematic tendency to assign higher methodological ratings to the studies confirming one's position? For example (to return to the dispute about psychotherapy outcome), would a pro-therapy group consistently assign higher ratings to the pro-therapy studies and a con-therapy group assign higher ratings to the con-therapy studies? Relative to the neutral judges, would each group assign higher ratings to the studies supporting their position and lower ratings to the studies supporting the opposing position? A comparison of the ratings of those holding opposing positions to the reliable ratings of the disinterested judges would provide a test of the potential distorting effects of an initial position (hypothesis) and confirmatory strategies as they are likely to be manifested in the process of conducting a literature review.

In other variations, judges with differing positions could be given a set of studies that were independently evaluated as confirmatory or disconfirmatory and simply asked to recall the percentage of confirmations and disconfirmations. Or, if hypothetical data were used, various dimensions of the data could be systematically manipulated, with the determination of standing along these dimensions established by neutral judges. For example, subtle or not so subtle flaws in research design or analysis could be introduced; methodological rigor could be varied; or the proportion of confirmatory, disconfirmatory, and neutral studies could be systematically controlled. Scientists' interpretations of these data would be analyzed relative to their positions and problem-solving strategies, providing an experimental means for evaluating generalization of Chapman and Chapman's findings. If parallel instances of illusory correlations and associated distortions were found to occur with scientists, which appears likely, it would provide scientific grounds for challenging the use of confirmatory strategies.

Mahoney's 1977 investigation addressed several of these issues. Rather than studying the processes involved in the review or evaluation of many related studies, however, he examined the way in which scientists review single articles that are being considered for publication. Mahoney constructed variations of a hypothetical research paper. Of interest here are two of the variations, one of which reported results supportive of the theoretical positions of reviewers and one of which reported nonsupportive results. The papers were otherwise identical, as in the sections describing methodology. The hypothetical research papers were sent to a series of reviewers who were asked to make recommendations about whether to accept the presumably real article for publication. They were also asked to rate the study along several dimensions, including its methodological adequacy. The reviewers who received the supportive variation of the paper showed a strong tendency to recommend publication and to assign it high methodological ratings, whereas those receiving the nonsupportive variation showed a strong tendency to recommend rejection and to assign it low methodological ratings. It appeared, then, that those receiving confirmatory data showed a general tendency to evaluate the research positively and that those receiving disconfirmatory (or non-supportive) data showed a general tendency to evaluate the research negatively, thereby providing a possible demonstration of problems arising from confirmatory judgment strategies.

The study of confirmatory strategies need not be limited to the review article or to evaluations by scientists of single research papers. More general study may provide meaningful insights into other scientific processes. For example, is it true that confirmatory strategies are closely related to scientific closed-mindedness? Assume that this is the case and that, as Kuhn argues, science is normally characterized by closed-mindedness and, thereby, confirmatory strategies. One could then explore the conditions required to create a shift to disconfirmatory strategies. One could also explore the extent to which the scientific method discourages confirmatory strategies and whether there are circumstances, other than the discovery of anomalies, in which adherence to the rules of scientific play override the tendency to use confirmatory strategies. Along other lines, one could ask whether problem-solving strategies are related to other variables, such as social factors. For example, does group cohesion foster or inhibit the use of disconfirmatory strategies? What type of social networks are most likely to encourage the use of disconfirmatory strategies? How does organizational hierarchy, level of collaborative effort, and work setting relate to the strategy used? One could also study deeper and more involved issues, such as the potential conditions under which confirmatory or disconfirmatory strategies are of greater usefulness, and the relationship this might hold to the developmental status of a scientific field.

Finally, deeper metaphysical issues can be raised. How might the variability inherent in the subject matter of a scientific field interact with problem-solving strategy, scientific closed-mindedness, and change in belief? Everyone but the most strictly deterministic of us assumes that the phenomena we observe across all areas of science contain a certain level of variation, which may be a combined product of random variation and nonrandom changes that the subject matter exhibits while still maintaining its identity.[6] The level of variation evidenced in the subject matter, however, may differ across scientific fields. Other things being equal, the level of variation and the level of consistency obtained among scientific observations of that subject matter should be inversely related. For example, with lesser levels of variation, consistency should be greater.

Suppose that there is a systematic relationship between the consistency of a demonstrated anomaly and its impact on belief systems. The more consistently the anomaly is demonstrated, the more likely it is to result in a change in belief. This relationship could interact with problem-solving strategy, in that the level of consistency with which anomalies must be demonstrated before belief changes increases to the extent that confirmatory strategies are used and closed-minded beliefs are held. If a rather strict confirmatory strategy is being used, scientists are unlikely to be swayed by the demonstration of an anomaly in, say, one experiment in a series of ten. In contrast, if the anomalous finding appears one hundred times in a series of one hundred experiments, even the most extreme of the closed-minded confirmatory strategists are likely to take note.

In summary, it is possible that the variation in subject matter can differ, that the variation in subject matter is systematically related to the consistency with which anomalies can be demonstrated, and that the latter interacts with problem-solving strategy to determine the likelihood that a belief will change. Therefore, all other things being equal, in fields with subject matter that displays greater variablility there is a decreased likelihood of uncovering consistent anomalous findings and, by implication, of achieving persuasive disconformations. The level of variability in the subject matter, then, may set a limit on the consistency of results that can be obtained. Unlike limits in consistency resulting from methodological factors, this source of variability is not a measurement artifact that can be overcome by more sophisticated technology. Rather, it is an inherent quality of the subject matter.

The potential significance of these statements can be understood better through application to an area of scientific inquiry. Take, for example, research in the social sciences. Closed-mindedness is seemingly common in the social sciences, and the level of variation in the subject matter apparently is high. Therefore, the likelihood of producing highly consistent demonstrations of an anomaly is low, reducing the chances of obtaining experimental results that provide convincing evidence against *any* position. One must overcome not only confirmatory strategies but also powerful factors operating against the discovery of consistent experimental results. These obstacles would almost invariably lead to a proliferation of differing positions because new ones would constantly be generated but few would die—be "disproved."

In contrast, Campbell (1978) has presented instances from the field of astronomy in which a critical experiment produced reliable findings that led to the repudiation of certain positions. But how likely is such an occurrence in the social sciences, given the possible existence of closed-mindedness and high variability in the subject matter? Given the extreme difficulties involved in achieving consistent findings in the social sciences, which may partially result from the subject matter's inherent variability, are major changes in investigative and evaluative strategies required and what modifications should be made? Further, will these factors operate against the discovery of exemplars in the social sciences, perhaps a necessary condition for mature scientific status? Exemplars usually seem to require effects that can be demonstrated with extreme regularity.

JUDGMENT ERRORS STEMMING FROM COGNITIVE LIMITATIONS

Cognitive limitations are almost certainly the most basic, most prevalent, and most troublesome source of human judgment difficulties. Many of the so-called bad judgment habits described above, such as overreliance on confirmatory strategies and underutilization of base rates, may be secondary products of cognitive limitations. These bad habits can be viewed as judgment strategies or as approaches individuals use when attempting to perform judgment tasks. Simon (1957) argues that because such strategies simplify the cognitive operations required for reaching judgments, they are used when we face reasoning tasks that outstrip our cognitive capacities. The judgment strategy assumed to underlie belief in the law of small numbers provides one example. Judging a sample's representativeness by evaluating the similarity between the characteristics of the sample and the assumed characteristics of the population bypasses the evaluation of other factors, such as sample size, and simplifies the required cognitive operations. Many bad judgment habits, then, may be accurately characterized as reasoning strategies that offer a way to compensate for our cognitive limitations. In many everyday situations, these strategies produce solutions that are satisfactory for our purposes. In other situations, they may lack the necessary precision or be extended to problems for which they are not suited, leading to major judgment errors.

Moreover, social checks and the application of relatively straightforward decision aids may reduce or counteract the potential errors resulting from bad habits; they may be minimally effective, however, in mitigating the problems caused by cognitive limitations.[7] When a prior hypothesis leads a scientist to assume that nonsupportive data offer confirmatory evidence for his or her position, a social check may be provided when another scientist, with a differing prior hypothesis, reanalyzes the original data and discovers the first scientist's error. Relatively straightforward decision aids might also be applied as checks against this or other bad habits, such as when one conducts correlational analysis of signs and symptoms to determine if the belief in certain relationships is illusory. Those who study and practice science have long recognized the presence of judgment errors, and numerous checks and controls for such errors have gradually been incorporated into most scientific fields. Contrary to popular opinion, these errors may often stem from bad habits and not from emotional biases. Nevertheless, many of these checks and controls are effective in counteracting the judgment errors stemming from this former source.

There has been, however, little recognition of the impact of cognitive limitations, and perhaps this partially explains why very few procedures or methods have been developed to mitigate the judgment errors stemming from them. Therefore, although more rigorous application of scientific methodology may counteract bad habits, it does little (if anything) to counteract cognitive limitations. And, as I will argue, social processes often provide little aid in overcoming cognitive limitations of the type we are considering. Cognitive limitations, then, may not only be the most pervasive cause of judgment errors, but they may also necessitate the development of entirely new procedures and methods to supplement current scientific practices.

The Cognitive Bounds of Groups versus Individuals

Some researchers have argued that group processes or social factors overcome the cognitive limitations of individual scientists. If this commonly occurs, then one raises only a pseudoissue when questioning the potential impact of individual scientists' cognitive limitations.

Therefore, before discussing descriptive and prescriptive implications of cognitive limitations, I will critically examine the relative cognitive capacities of the individual and the group.

Of specific interest here is an individual's capacity to comprehend complex relationships. In most cases, in order to grasp the meaning of complex relationships, or the whole, one must be able to consider simultaneously the parts of the whole and their interrelationships — one must keep in mind, either simultaneously or in close temporal proximity, the parts and their interrelationships. This particular cognitive act should not be confused with the manner in which information is presented. The two are largely independent. Pieces of information need not be *presented* in close temporal proximity (although different rates and forms of presentation may facilitate or hinder comprehension), and different pieces can even be presented years apart. But to *comprehend* complex interrelationships, or the whole, at some point one must simultaneously hold these pieces in mind. Limits in the ability to do so set a ceiling on the complexity of problems that can be grasped.

This principle has been repeatedly emphasized and documented by Luria ([1962] 1980). Although Luria has generally worked with brain-damaged patients, their performance on complex problems lucidly illustrates the relationship between limitations in the capacity to simultaneously hold information in mind and limitations in the ability to solve complex problems. Patients with damage to a particular area of the left parietal lobe often evidence a specific type of language disorder. They retain the capacity to understand component parts of speech and comprehend abstract verbal concepts; for example, they can understand and define single words and can grasp the relationship between two abstract verbal concepts, such as liberty and justice. These patients can also comprehend many phrases and sentences adequately, but they cannot understand a statement such as the following: "The women who worked at the factory came to the school where Dora studied to give a talk" (p. 504). The patients understand the individual words and even groups of words within the sentence, but they cannot interpret the overall meaning of the construction, often mistaking such things as who gave the talk and where the talk was given.

In an attempt to decipher the meaning of this sentence these patients may break it into fragments, but they are still unable to piece the sentence together. To understand such a sentence, one must co-ordinate the details into a common whole, which requires the simultaneous grasp of its components. These patients simply cannot perform this latter function, precluding their comprehension of the interrelationships of the component parts and, therefore, the meaning of the entire sentence. Luria argues that it is precisely this inability to keep language components and their interrelations in mind simultaneously that accounts for these patients' general deficit—an inability to comprehend complex grammatical relationships, such as those contained in the above sentence.

There is little that another individual or a group can do to aid these patients' comprehension of grammatically complex relationships.[8] Direct explanation of the meaning of the sentence provides no help because the cognitive operations needed to understand the explanation equal those needed to understand the sentence. In contrast to many types of problems, no direct explanations simplify the cognitive operations needed to solve problems that exceed one's ability to grasp interrelationships.[9] Given this, even if the patient is offered a number of alternative explanations, one of which is correct, he or she cannot recognize the correct alternative. Again, the cognitive operations required to grasp alternative explanations (assuming the complexity of the incorrect alternatives equals that of the correct alternative) are no less complex than those required to grasp the original sentence. Additionally, if direct explanations do not help, it is almost inconceivable that a group of such brain-damaged individuals could work together to obtain a comprehension of this or equally complex sentences. No contribution of one or more patients to other patients would increase any other patient's ability to comprehend the overall meaning of such sentences. For example, one patient might comprehend a word or a part of the sentence that a second patient did not, but this would not aid in this second patient's grasp of the sentence's overall meaning. If, by chance alone, patients could derive correct answers to questions about these sentences, no one would have the capacity to understand why these answers were correct, much less what the sentences meant.

This example is not intended to cast aspersions on scientists, but it is used to make a point. Only relative capacity separates scientists from these patients. The *relationship* that exists between these patients' capacity to hold information simultaneously in mind and their capacity to solve complex problems does not set them apart, for this relationship exists for all individuals. It is rather that both their capacity to hold information in mind and to solve complex problems are pathologically limited. The scientist's capacity to perform complex cognitive operations is certainly greater, but it is certainly not unlimited.

There is little question that the individual scientist's limitations in the ability to hold information simultaneously in mind limits his or her capacity to solve problems requiring the discovery or identification of interrelationships among data. If one can comprehend three-way but not four-way interactions, then it might be possible to identify three-way interactions that exist within data or in nature, but it is highly unlikely that one will identify four-way interactions. Although, by some form of groping, one might identify or map out a four-way interaction, one would still be unlikely to comprehend it. One's capacity to use knowledge of this four-way interaction either to design tests to delineate this phenomenon further or to construct a more comprehensive vision of nature would be even more limited. Clearly, then, as the complexity of patterns within data increases beyond the individual scientist's cognitive limitations, the likelihood of uncovering these patterns decreases proportionately. Further, the scientist's ability to comprehend what has been "identified" and to use it as a basis for conducting further intelligible tests or for generating more comprehensive theoretical constructions also decreases proportionately.

There appears to be no means by which the joint working of the scientific community can directly create comprehension abilities that extend the limitations of individual minds to grasp complex configural relationships. The community is limited by what the most complex mind can grasp. This is akin to the problems faced by biologists before the discovery of the microscope. Everyone was limited by what the best pair of eyes could see, and everyone looking together could provide no finer resolution. The above is *not* to say that the combined efforts of a community do not make better puzzle-solving possible. One person may discover what a second person does not, making combined efforts to identify patterns in data more successful than individual efforts. But another individual, or the combined working of individuals, does not make possible the comprehension of patterns at a level of complexity beyond that which the single most complicated mind can grasp. The complexity of what can be grasped is not increased if there are many minds working on a problem because all minds are limited by what each can grasp. Expressed in information-processing terms, each individual has limited channel capacity, which cannot be increased by combined effort. Even if scientists could in some way pool ideas or minds such that more complex patterns were uncovered, no individual scientist would be able to understand what was uncovered. In the comprehension of complex relationships, the capacity of the whole (community) is no greater than the capacity of the individual parts.

How, then, has it been possible for scientists to derive complex puzzle solutions? Are there mechanisms through which cognitive capacities can be increased, and how might the community contribute to this accomplishment? In addressing these difficult questions, one can start by critically examining underlying assumptions about the typical complexity of scientific solutions. Many great puzzle solutions or scientific insights, at least in the hard sciences, have not been distinguished by their delineation of complex interrelationships. Rather, their parsimony and simplicity are especially striking. For example, Kepler's laws of planetary motion and Newton's theory provided surprisingly simple solutions to perplexing problems. Newton himself said that "truth is ever to be found in simplicity . . . ," and Einstein described himself as "one who seeks only the trustworthy source of truth in mathematical simplicity."[10] Many of Einstein's greatest scientific insights are characterized by an elegant simplicity. If these examples are representative, we may have overestimated the complexity of typical scientific solutions. Although the available data are complex, the identified interrelationships are relatively simple. False assumptions about the complexity of the scientific insights that ultimately find strong empirical support may be based on attention to the complexity of available data. Direct attention to the complexity of solutions might lead to a more accurate appraisal. If relative simplicity in puzzle solution is the rule, many cases of supposedly complex solutions do not have to be reconciled with statements about limitations in cognitive complexity. On the contrary, these solutions were not complex.

Transcending Cognitive Limitations

Few cases of compex solutions may exist, but we must account for those that do. Under certain conditions (as alluded to previously), it is possible to transcend individual cognitive limitations, and the community often makes an essential contribution to this achievement. At least two of these conditions can be identified: when one integrates descriptions of component parts of a phenomenon that have been studied in isolation, and when one can transform data in a way that reduces the processing load on the individual.

First, consistent with analytic tenets, one may be able to break a complex phenomenon into component parts and to study these parts in isolation without irreparably distorting the way they operate when combined. Once some scheme or framework for describing and predicting the operation of isolated parts is achieved, it may then be possible to evolve a description of the operation of two or more parts in combination. Since the scheme for describing or predicting the operation of individual parts may reach the upper bounds of human cognitive complexity, that for describing or predicting the combined working of parts may possess a level of complexity that exceeds individual cognitive capacities. This complex proposed solution is reached by trying to combine the individual descriptions of the parts, a process that may be characterized more by groping than by systematic and orderly effort. The construction of such complex attempts at puzzle solving is probably most likely when unifying themes or constructs cannot be identified and when previously successful explanatory principles or mathematical axioms are readily available and, roughly speaking, can be added on to the descriptions that have already been evolved. Clark Hull's (1943, 1952) attempts to develop a theory to predict behavior, which involved the construction of an increasing complex mathematical formula, may provide one example of this process.

The proposed puzzle solution, having acquired greater complexity than the framework pertaining to individual parts, may not be fully comprehended by individuals in the scientific community. Nevertheless, it may be useful as a potential problem solution so long as tests of the construction can be conducted. If experimental results are deemed supportive, members of the community may conclude that the combined puzzle solution grasps something essential about nature; they will likely try to refine this solution or add further pieces to it. But there is a cost in attempts to describe or predict the working of combined parts with puzzle solutions at a level of complexity beyond the comprehension of individual scientists. Much plodding and many trial-and-error efforts are usually necessary before a solution that is even temporarily satisfactory can be found. And attempts to use this solution as a basis for building more complex ones are far more difficult and inefficient, and far less likely to succeed.

There is a second potential condition under which individual limitations can be transcended and complex solutions thereby attained. When data are transformed in a way that reduces the processing load on the individual, one can comprehend more information simultaneously. Transformation of this type can occur by developing a system for representing or expressing data that condenses information and allows more to be taken into account simultaneously. This is often accomplished through the development of a more efficient language for describing the scientist's relevant body of knowledge. For example, translating verbal descriptions to mathematical formulae may accomplish condensation. Language systems, however, are only one of the means available. Graphic display of data using computer assistance is another. Transformation can also take place internally, rather than through the development of alternate means for expressing information. By overlearning material, that is, by mastering it to the point that its recall or utilization requires minimal conscious effort, one can reduce the "channel capacity" required for its use (Allport, Antonis & Reynolds 1972; Kahneman 1973; Posner 1973; Shaffer 1974). This is analogous to the manner in which a pianist learns to play a different melody simultaneously with each hand. Although it may be impossible initially to play the melodies simultaneously, once the material for each hand is thoroughly mastered, they can then be played together. Similarly, by overlearning certain information scientists may be able to reduce the channel capacity required for its utilization. The freed channel capacity would permit the scientist to simultaneously keep in mind additional information, resulting in a greater total than was previously possible.[11] This certainly occurs with students as they become more familiar with material in their field, but it is uncertain whether it also occurs once scientific expertise has been acquired. To my knowledge, there has yet to be a single study with scientists that has directly addressed this possibility.

If scientists can and do obtain increased capacities to hold information in mind through overmastery, it is unclear what additional cognitive abilities are realized. The possibilities range from the capacity to consider one or a few more variables in a linear fashion to the capacity to grasp complex interactions and interrelationships.

For example, we do not know whether an increased capacity to hold information in mind simultaneously produces an increased capacity to comprehend *configural* relationships. Increased capacity is likely a necessary but not sufficient condition for the latter. Nor do we know the extent to which expansion of overmastery can continue to produce increments in the capacity to hold information in mind.

How might the community or social processes contribute to the means by which individual scientists solve puzzles and at times transcend their cognitive limitations? Many puzzle solutions may not reflect complex cognitive operations, and in such cases the community does not and need not contribute by increasing cognitive complexity because it is simply unnecessary. Rather, the community can contribute through its sheer number, providing an ensemble of scientists to work on the same or related problems. This increases the probability of obtaining a proposed solution that a significant proportion of the community endorses as promising or fruitful.

Under conditions in which a proposed solution extends beyond the cognitive capacities of individual scientists, the community can contribute, for one, by providing a system for expressing or representing data in a way that condenses information. The generation of such a system usually does not require thinking that surpasses individual limitations in the capacity to hold information simultaneously in mind. Therefore, group effort increases the likelihood that someone will develop a useful system. Of greater interest, however, is the condition under which pieces of nature have been studied in isolation; subsequently, proposed solutions for interrelating these parts exceed the cognitive capacity of individual scientists. Under these conditions, group effort makes possible the generation and evaluation of many potential puzzle solutions or of ways to combine knowledge of individual parts. The puzzle solutions may result in combinations of pieces or parts that are beyond the cognitive capacities of the individual scientist or the combined minds of the community, so the search for the "correct" combination must in some sense be blind. Therefore, the greater the number of combinations that can be derived and tested, the greater the probability of finding a correct solution. This process is similar to the advantage held by a species with a large and frequently reproducing population. A successful permutation is more likely to occur and be "selected."[12]

The population of scientists, then, contributes to the construction of a problem solution that is beyond the cognitive capacities of the individual members by generating and testing far more permutations than would be possible by individual effort. Although this effort may be largely blind and fortuitous, group effort does increase the probability of achieving success. Ironically, the community may generate a successful solution that, as a whole, extends beyond the cognitive capacities of the individual scientists and cannot be fully comprehended by any individual. But full comprehension is not necessary for selection, only some way of testing the solution. Under conditions requiring the construction of solutions beyond the cognitive complexity of individuals or the community, the whole is greater than the capacity of each part, not because a combination of minds allows for the comprehension of greater complexity, but because a combination of individual effort makes the generation of a successful permutation more likely. When such a solution is obtained, intense emersion in the world construction may trigger such processes as overlearning and other mechanisms for transcending individual limitations, which can eventually make the new world construction comprehensible. In other cases, comprehension may never be forthcoming.

In summary, the community contributes by increasing the number of solutions that can be generated and tested, thereby increasing the probability of obtaining a successful solution that is beyond the cognitive capacity of individual scientists. But the community does not contribute by directly increasing the cognitive capacity of individuals to grasp complex interrelationships—within this realm the cognitive capacity of the combined community is no greater than the cognitive capacity of the individual parts (scientists). Therefore, the potential impact of the individual scientist's cognitive limitations is not a pseudoissue.

My argument pertains to restrictions in the capacity of individuals to grasp complex configural relationships, an exceedingly important type of cognitive limitation. Although I have emphasized boundaries on the extent to which the community can aid in overcoming these limitations, there is little doubt that the community can effectively aid in overcoming other types of cognitive limitations and perform functions that are beyond the capabilities of individuals working in isolation. The community, for example, can be very effective in overcoming limitations in the amount of information that single individuals can acquire. An individual physician cannot possibly acquire the great bulk of relevant medical information. But when he or she is confronted with an unfamiliar set of symptoms, there is often easy access to a community of specialists who may possess the required knowledge. Parallel instances abound in the basic sciences, one being the communication that takes place among members of the "invisible colleges," or groups of leading scientists sharing related interests, that Crane (1972) has described. Hayek (1955) provides an illuminating and thoughtful discussion of the manner in which, and the mechanisms by which, groups can realize accomplishments that isolated individuals or single minds cannot.

RECONCILING DESCRIPTIONS OF SCIENCE WITH THE COGNITIVE LIMITATIONS OF SCIENTISTS

If we grant the possibility that findings pertaining to individual cognitive limitations generalize in a manner relevant to individual scientists and scientific groups, what implications does this possibility hold? What impact might the generalizations of the judgment findings have on descriptions of science?

The recognition of bad judgment habits and of failures to use normative guidelines is unlikely to substantially alter our view of science. It may lead to an awareness of new sources and forms of judgment error, but these errors largely overlap with those previously identified as stemming from bias. As such, the documentation of errors based on bad habits and failures to use normative methods, more or less, adds to an already well-established list. But recognition of cognitive limitations may alter our view of science profoundly. There appears to be little appreciation that many judgment errors and nonoptimal performances may result not from doing the wrong thing, but from an incapacity to do the right thing. Although the effects of bad habits and failures to use normative guidelines can be attenuated by scrupulous application of available decision aids and conventional rules of scientific procedure, there exist few such forms of assistance that might counteract the consequences of cognitive limitations. In addition, one can potentially eliminate bad judgment habits, but not cognitive limitations; one can only attempt to compensate for them. Finally, the general failure to appreciate cognitive limitations seemingly underlies the design of prescriptive programs that overestimate human capacity and, therefore, make unrealistic demands upon scientists. For these reasons, recognition of cognitive limitations may substantially alter our beliefs about how scientists do and can function, leading in turn to major revisions in prescriptive programs.

Assuming for the moment that judgment findings pertaining to cognitive limitations do generalize, what image (description) of the scientist might emerge? As I argued in chapter 1, complex considerations and information are relevant to most, if not all, scientific judgments. Across all scientific activities and so-called levels of inference, from the "seeing" of facts to the "seeing" of theories, from the measurement of atomistic units to the measurement of hypothetical constructs, from the selection of laboratory equipment to the selection of research strategy, complexity is the rule. The judgment literature suggests that individuals are least able to perform adequately exactly *these* types of decisions—those that require the cognitive management of considerable complexity.

Therefore, the generalization of findings to scientists suggests all of the following major possibilities. *First*, the absolute level of performance on many scientific judgment tasks is frequently low. In general, the reliability, much less the validity, of scientific judgments is far poorer than we have generally assumed. *Second*, when making decisions, scientists frequently attend to only a few variables or cues (although they may believe otherwise). Once they are given access to a limited amount of information, additional information generally does not increase their judgment accuracy. It is not that additional information is uninformative, but that scientists cannot select the most useful information, weight it appropriately, or integrate it adequately. *Third*, scientists may have sufficient cognitive ability to comprehend simple configural relationships among cues or variables, but insufficient ability to comprehend more complex relationships. Although they may be able to construct models or theories of nature that describe more complex configural relationships, these constructions provide little or no increment in judgment performance relative to the level that can be attained by simply adding variables in a linear fashion. *Fourth*, it is possible to construct models that accurately reproduce scientists' judgments. Even simple linear or nonconfigural models can probably account for the great majority of scientists' judgment variance. *Fifth*, and perhaps most important, scientists' level of judgment performance can be surpassed by actuarial methods.

As these hypotheses suggest, generalization of the judgment findings to science would call for a considerable shift in viewpoint. We would probably move towards a general description of the scientist as a limited being, one who is incapable of satisfying many of our scientific ideals and far less capable than we have generally assumed of managing complex problems. This shift in viewpoint might best be captured by reworking a well-known statement by Descartes that reflects what is now a common and rather optimistic viewpoint about human potential—that is only the incorrect application of the mind that restricts the growth of knowledge. Descartes said, "To be possessed of a rigorous mind is not enough: the prime requisite is to apply it rightly" ([1637] 1912). The new description might rather be: "To be possessed of a rigorous mind and to apply it rightly is not enough: the prime requisite is to be possessed of a more rigorous (powerful) mind."

Implausible Descriptions

I do not know whether these five hypotheses, taken alone or in combination, overstate the case for cognitive limitations as they are manifested in science. Regardless, granting the plausibility of the overriding point—that, like all others, scientists are limited cognitive beings with relatively harsh restrictions in their capacity to perform complex judgment tasks—we can see whether popular descriptions of science fit within the boundaries of human ability as outlined by the judgment literature. We have only a rough idea about what these boundaries might be, although in certain cases it is clear that humanly impossible cognitive acts have been described. One would hope that the psychology of science will eventually provide a far more precise picture of human cognitive limitations, one that will directly aid in evaluating the plausibility of descriptions.

The descriptive implications of cognitive limits are seemingly most relevant to the historical study of science, but they also relate to the increasingly popular practice of studying scientists at work (for examples, see the work of Latour & Woolgar 1979; Lynch 1981/Mitroff 1974). The descriptive study of historical and current scientific events is itself theory laden, and the interpretations and conclusions reached by investigators are shaped by prior assumptions. Inaccurate assumptions can lead to inaccurate descriptions, and individuals describing historical and current occurrences often seem to overestimate the cognitive capacities of scientists. This may explain the tendency to describe judgments and decision-making processes that are probably beyond the cognitive capacities of scientists.

Let me start with a broad example and then move to a series of specific ones. Kuhn (1970) assumes that scientists differentially apply a set of values when evaluating a theory or paradigm. Kuhn does not address cognitive strategies explicitly, but there are indications that he assumes these evaluations entail the use of complex configural judgments. If Kuhn does hold this position, he may have assumed cognitive capacities that are beyond that of scientists. Therefore, not only his assumption about the evaluation of paradigms is suspect, but any interdependent assumptions as well.

The situations in which complex and rather remarkable cognitive capacities have been assumed or described are not limited to one or a few areas, such as the evaluation of paradigms, but include a wide range of scientific activities. A book by W. I. B. Beveridge (1957) can be used to provide specific examples of questionable assumptions and descriptions. Without intending to select out Beveridge's work, I find that it is suitable to such an examination. It is the product of an intelligent and widely read man, offers broad coverage of scientific activities, endeavors to describe mental activities that are common to science, and contains many statements that are typical of those found in writings on science.

In chapter 1 we divided science into the facets of information gathering, information evaluation, and assumptions. Beveridge's statements about these first two facets will be examined in turn and contrasted to what is known, or at least suspected, about human capabilities. In the area of information gathering, the judgment literature suggests that scientists are able to integrate or consider only a relatively limited range of information. In discussing this area, however, Beveridge makes statements like: "The careful recording of all details in experimental work is an elementary but important rule" (p. 24). "We need to train our powers of observation . . . and make a habit of examining every clue that chance presents" (p. 44). "The successful scientist gives attention to every unexpected happening or observation that chance offers . . ." (p. 45). Considering the very limited number of cues to which individuals seem able to attend and the immense complexity of the sensory world, what is the likelihood that these descriptive statements are accurate?

In the area of information evaluation (which admittedly overlaps substantially with information gathering), the judgment literature suggests that individuals can simultaneously keep in mind only a limited amount of information and have great difficulty comprehending or identifying complex interrelationships among data. Compare this assumed state of affairs with Beveridge's comments: "When the experiment is complete and the results have been assessed . . . they are interpreted by relating them to all that is already known about the subject" (p. 27). "Having arrived at a clear understanding . . . at every step in our reasoning it is essential to pause and consider whether all conceivable alternatives have been taken into account" (p. 116). "As many hypotheses as possible should be invented and all kept in mind during the investigation" (p. 68). Beveridge's last statement, while prescriptive, implies that scientists can and do perform such cognitive feats. But what is the likelihood that scientists can perform these or any of the other activities described in the above quotations?

When taken in the context of findings on human cognitive limitations, Beveridge's descriptions may seem almost absurd. Although the judgment research currently allows one to identify only rough boundary conditions for evaluating descriptive plausibility, there is little question that Beveridge has violated these boundaries by a wide margin. But when compared with claims made by some writers, Beveridge's statements seem moderate. By no means are his ideas out of line with many descriptions of science, descriptions that are finally coming under increased scrutiny. Nevertheless, in describing historical and current activities in science, investigators continue to assume, uncritically and perhaps mistakenly, rather remarkable cognitive capacities. Any description of scientific processes assuming or reporting complex judgment capacities must be critically examined. The impossible may have been posited.

Possible Reinterpretations: Theory Evaluation and Discovery

Thus far we have addressed rather broad issues. To summarize, I suggested that the generalization of findings on cognitive limitations would necessitate changes in descriptions of science. I presented examples from the works of two authors to support the argument that some, if not many, descriptions of science suppose or detail cognitive activities that scientists cannot possibly perform. If this is the case, then substantial portions of the literature describing science may have to be altered to incorporate the realization that scientists are cognitively limited beings. This major shift in viewpoint would not only result in the reinterpretation of numerous historical events and everyday scientific activities, but would also have important prescriptive implications. Examples of possible reinterpretations will now be outlined before discussing prescriptive implications.

Relevance to Theory Evaluation

Theory evaluation provides an initial example of scientific practice that may require reinterpretation if judgment findings generalize. The task is complex: data do not speak for themselves, and even strategies for theory evaluation that in principle appear straightforward, such as disconfirmation, in practice leave intricate problems unsolved and do not circumvent complexity. Data from multiple experiments must be evaluated. One must develop strategies for delineating the boundaries between relevant and irrelevant studies, weighting various findings, and integrating studies that may use different methodologies, terminologies, and viewpoints. Not only must evaluation be conducted of data-based and related criteria, but one must also consider criteria that are not strictly data-based, such as the theory's parsimony, its logical consistency, and even its promise for generating fruitful "puzzles." These criteria should be evaluated configuratively and not by simply adding, greatly increasing the complexity of this judgment task. And there are few objective procedures that might aid the scientist in making these complex judgments, nor even systematic or generally accepted procedures for doing so. Rather, given the rare case in which one or a few studies conclusively result in the rejection of a theory, theory evaluation requires subjective (although *not* necessarily irrational) and extremely complex judgments.

These are the judgment requirements the scientist-theoretician is expected to meet when evaluating theories. Given the consistent finding that individuals cannot adequately perform far less complex judgments, what is the possibility that scientists can adequately perform the exceedingly complex judgments required for theory evaluation? It seems likely, then, that judgment findings do generalize to theory evaluation and that such evaluations are performed far less reliably and validly than is generally assumed. Instead of successfully integrating complex networks of information and considerations, in practice scientists may heavily weight a very limited set of information and virtually ignore the remaining relevant information. Simple configural relationships among variables might be recognized, but more complex relationships are probably misinterpreted or not even detected. Scientists may, nevertheless, hold the mistaken impression that they have allotted the appropriate weights to a wide range of information, have integrated a large amount of information, and have accurately detected and utilized complex configural relationships.

This reinterpretation of theory evaluation touches upon even deeper considerations. If virtually all prescriptive programs for theory evaluation require complex decisions, it would follow that—regardless of scientists' intentions—*few, if any, of the popular philosophical programs for theory evaluation can be performed as required*, given current methods. If this is the case, successful episodes in the history of science cannot be strictly attributed to the implementation of any such prescriptive programs. If they, in part, require what individuals cannot possibly do, they, in part, cannot possibly have been performed. Therefore, their strict implementation cannot account for scientific success. This casts doubt upon many popular attempts to reconstruct (describe) successful scientific efforts as products of conformity to a set of prescriptive guidelines. (For an excellent overview of historiographical attempts to reconstruct and explain science, see Weimer 1974a, 1974b.) Exactly how scientists have proceeded, the extent to which these procedures resemble various prescriptive programs, and the elements of procedure that account for scientific success would be open to question. Further, if prescriptions do require the impossible, it is reasonable to argue that they should be revised. Possible directions for revision, such as the development of actuarial methods for theory evaluation, will be discussed under a later heading.

Relevance to the Context of Discovery

Although the boundaries between the contexts of evaluation (justification) and discovery are considerably blurred, I ask the reader to treat the two as distinct for the sake of the following discussion. Might research findings pertaining to complex judgment and cognitive limitations also apply to the context of discovery? Attempts at scientific discovery often involve conditions in which the scientist is required to interpret or comprehend a complex pattern of available information. Many judgment studies have been directed towards analogous conditions, in which individuals are required to assess or interpret complex information. These studies cover a broad spectrum. In some, information must be coded or classified into predefined categories ranging from those that are rigidly defined (Hislop & Brooks 1968; Levine 1969) to those that are relatively open (Posner & Keele 1968; Vygotsky 1962). In others, individuals attempt to uncover the organizing principles (concepts) or configural relationships inherent in the data (Restle & Brown 1970). Individuals often perform poorly on these tasks, especially when asked to discover or identify complex concepts or interrelationships. Scientists trying to identify or discover complex relationships or to organize a series of discoveries, as theory construction requires, are faced with an analogous but considerably more demanding task. Is there any reason to assume that scientists do far better when attempting to comprehend and organize a complex series of relationships, given that the scientific method provides limited aids for such accomplishments? If findings demonstrating the limited abilities of individuals to identify and comprehend complex relationships generalize to scientists, one would hypothesize that scientists' performances on such tasks are often far from optimal and subject to marked restrictions in managing complex information.

Reinterpreting Other Scientific Practices and Common Occurrences

When discussing possible reinterpretations of theory evaluation and discovery, the level of performance on these tasks and related implications were given primary consideration. Judgments across a wide range of scientific activities may also be characterized by surprisingly low performance levels, but there is little purpose in listing these activities and reiterating the points made in reference to theory evaluation and discovery. It is more useful to present examples of how cognitive limitations may provide alternative accounts or explanations for common scientific occurrences and practices. Explanations that incorporate findings on cognitive limitations might initially change descriptions of these practices and occurrences only slightly; however, such an altered explanatory framework has different prescriptive implications. It is also likely to foster different investigatory strategies, which typically lead to different discoveries and, eventually, different descriptions.

Reasoning strategies and aids

Many of the reasoning strategies and aids used by scientists, ranging from themes and models for interpreting phenomena to prescriptive programs for theory evaluation, can be seen as procedures that reduce cognitive strain. These approaches narrow the range of information that must be considered, allow scientists to overlook inconsistencies, and serve to organize information, thereby simplifying decision making.[13] As such, many reasoning strategies and aids may stem from cognitive limitations and provide one means of compensating for them.

THEMES AND ORIENTATIONS

Scientific theories or assumptive networks are often believed to be very complex, but this belief may be based on attention to superficial appearance rather than underlying structure. There is some evidence that a limited set of relatively simple organizing schemes or core concepts underlies the apparent surface complexities of many assumptive networks. Two variations of such fundamental core concepts are themata and orientations.

Holton (1973, 1978) deserves major credit for describing the characteristics and functions of scientific themata. Themata are relatively simple organizing concepts that reflect basic assumptions about nature's characteristics and about preferred forms for expressing or constructing theoretical ideas. Among other things, themes help the scientist fit complex phenomena into a simpler and more familiar framework and, thereby, facilitate interpretation. Through the application of themes, what is complex and perplexing becomes comprehensible. Themata, then, simplify the task of interpretation and may represent one form of adaptation to cognitive limitations.

Orientations also serve as tools for interpreting nature, but they are more basic than themata.[14] Orientations are the core metaphysical assumptions upon which other metaphysical assumptions or frameworks (such as themata) are built. Metaphysical assumptions can be conceptualized as beliefs that are organized within a hierarchical structure, with orientations representing the most basic or fundamental level of the hierarchy. This abstract description can be clarified through discussion of two specific orientations and a study of scientific schools examining the two.

These two orientations, which will be referred to as the stimulus and symbolic orientations, are assumed to fall at the endpoints of a continuum. The orientations can be integrated, and intermediate points on the continuum represent varying levels of integration. The two orientations serve to organize a diversity of scientific ideas, including those pertaining to the methodology and knowledge base of a scientific school. Each can be described briefly. The stimulus orientation holds certain similarities to pragmatic perspectives. The meaning of events and the accuracy and utility of ideas are evaluated through reference to external world occurrences. Attention is directed to means-ends or causal relationships between variables. Ideas are considered valid if they accurately describe such means-ends relationships or provide the capacity to control them. Theories are viewed as means to ends, that is, as means to generating useful facts or accurate descriptions of causal relationships. As one might expect from this list of assumptions, the stimulus orientation is associated with positivistic, reductionistic, realistic, and strict deterministic beliefs.

The symbolic orientation differs on all of these points. The meaning and accuracy of ideas are evaluated against criteria with no direct external world referents. These criteria address the ideal form or structure of ideas, such as their parsimony and logical consistency. The extent to which ideas are assumed to achieve ideal form determines judgments about their validity. Theories are viewed as ends in and of themselves, with these ends representing the attainment of "truth" or knowledge. Therefore, facts are seen as a means for assessing the accuracy of theories. In general, the symbolic orientation is associated with nonpositivistic, holistic, idealistic, and nondeterministic beliefs.

The two orientations are, we have noted, endpoints on a continuum, and few scientists hold beliefs that fall at either endpoint. Rather, most have integrated the orientations to varying degrees and fall at some intermediate point. Moreover, scientists within the same school or community often share similar positions on the stimulus-symbolic continuum. Less mature schools hold more extreme positions or fall more towards the endpoints of the continuum, having integrated the two orientations to a lesser extent than have more mature schools.

Several of these assumptions were tested in a study of the behavior modification and cognitive development groups in psychology, which are major subcommunities of the behavioral and cognitive schools,

respectively. These subcommunities had been independently identified as prototypes of the relatively extreme stimulus orientation (the
behaviorists) and the relatively extreme symbolic orientation (the
cognitivists). A questionnaire assessing a broad spectrum of beliefs,
ranging from views on the continuity between animal and human behavior to views on determinism, was sent to members of the subcommunities. If these subcommunities had been accurately identified as
holding relatively extreme viewpoints, one would expect that the
responses of the behavioral group to the questionnaire would consistently endorse stimulus beliefs and responses of the cognitive group
would endorse symbolic beliefs. Therefore, one factor—standing on
the stimulus-symbolic continuum—should account for the pattern of
beliefs exhibited within the schools and the differences in belief exhibited across the schools.

It was predicted that factor analysis of the responses would produce a one-factor solution to which all sampled areas of belief would
be closely related. In other words, mathematical analysis of the responses would show that they clustered into one unified group as
opposed to a set of groups. Given that the diverse beliefs assessed by
the questionnaire are not typically thought of as unified along a single
dimension, and given the rarity of obtaining one-factor solutions,
support for such a prediction seemed unlikely, at least at face value.
Nevertheless, a one-factor solution clearly emerged—responses did
line up along a single dimension—with the pattern of results supporting the conclusion that this factor represented the stimulus-symbolic
continuum. Further, all sampled areas of belief loaded heavily on this
one factor—they showed a close relation to overall position on the
stimulus-symbolic continuum. A scale was then derived to measure
the scientists' standing on the continuum. Analysis showed that the
behaviorists held a stimulus orientation and the cognitivists a symbolic
orientation.

These results indicate that standing on *one* underlying factor may
actually account for many of the assumedly multidimensional and
complex differences in belief exhibited by behavioral and cognitive
psychologists. The results also lend plausibility to the hypothesis that
the diversity of beliefs evidenced within and across scientific schools
may be a phenotypic manifestation of simple underlying genotypic
differences. Accordingly, one cannot uncritically assume that surface
complexities reflect underlying complexities. It is easy to imagine

various mechanisms by which simple cognitive strategies or decision rules could create surface complexity. For example, suppose objects differing across various dimensions are presented in pairs. In sorting or classifying each pair of objects, a simple rule is applied: Sort the larger of the two objects into group A and the smaller of the objects into group B. After a series of presentations, objects might appear together in each group that vary along multiple dimensions, but this diversity has been created by one decision rule—sort according to size. Similarly, orientations or themes may represent basic beliefs that underlie the selection or construction of less basic beliefs, the latter appearing to reflect underlying structures that are diverse and multidimensional. The study of human judgment leads one to hypothesize that scientists' cognitive limitations necessitate the use of relatively simple reasoning structures and that such structures provide an underlying unity to supposedly complex assumptive networks.

PRESCRIPTIVE PROGRAMS FOR EVALUATING AND CONSTRUCTING SCIENTIFIC REALITY

Like themes and orientations, prescriptive programs can also be viewed as problem-solving approaches that potentially reduce cognitive strain. Prescriptive programs provide simplifying strategies for decision making and facilitate the management of information, in part by allowing scientists to ignore inconsistent information and supposed irrelevancies. The utilization of specific programs by scientific communities may be closely related to the extent to which these programs reduce cognitive complexity and are believed to facilitate effective decision making. Scientists frequently reject prescriptive programs when they find that these approaches fail to solve or extend to problems considered too important to ignore. Apparently, such failures often occur because these programs are oversimplifications.

Three illustrative examples can be presented. Mill's Canons, once considered by some to be the total set of reasoning principles necessary to achieve scientific truth, depend on the assumption that all additional possible causes of a phenomenon are known and controlled. This assumption can rarely, if ever, be granted, and Mill's Canons offer no help in uncovering or controlling additional possible causes. Thus, the canons were found to be insufficient for solving the problems created by these and other complexities. Popper's falsification strategy provides in principle a simple method for evaluating theories. But it too has been found to be too simple. For example, it provides no guidelines for determining whether theory or method is at fault and what aspects of a theory require modification. Further, many argue that strict application of falsification strategy leads to premature rejection of theories and related practices that usually hinder scientific progress. Operationalism has also been found to be an oversimplification. It forces one to conclude, for example, that conceptually related measurement procedures that produce conceptually related outcomes measure different variables or constructs. All new measurement procedures must be considered new constructs, and there are no independent criteria for assessing the validity of these "constructs" or their possible relationship to other constructs. Like Mill's Canons and Popper's falsification strategy, operationalism is too simple and thus fails to provide decision rules that cover important considerations.

One need not possess complex cognitive abilities to discover instances in which relatively simple strategies do not solve relevant problems. Attempts to apply prescriptive strategies, either at the level of philosophical thought experiments or in actual scientific practice, inevitably expose their shortcomings and lead to their modification or elaboration. But such changes often make prescriptive programs more cumbersome to apply. (For a prime example, see Radnitzky's 1979 attempt to expand and improve upon Popper's methods.) Among other things, it may be this difficulty that leads scientists to reject or abandon these programs: they are no longer simple or easy to apply and can no longer reduce the complexity of scientific decision making. Regardless of whether difficulties in application underlie the rejection of prescriptive programs, one can still view them as simplifying strategies that potentially serve to compensate for cognitive limitations. Their use in scientific communities may be closely related to their effectiveness in reducing cognitive strain.

STRATEGIES OF INQUIRY AND OF EXPLANATION

Cognitive limitations may be associated with strong preferences for strategies of inquiry and explanation that simplify decision making. The analytic approach to science is possibly the most prominent example. Analytic methodology dictates that one study isolated phenomena and look for causal links between independent and dependent variables. Related preferences call for conducting highly specialized research and for strictly limiting the domain of inquiry. These methodological predilections typically reflect a series of underlying assumptions: that events are linked by linear causal relationships and that more complex interactive relationships are of secondary importance, or at least can be temporarily dismissed. Reductionism is often embraced, and more complex entities are believed to stem from less complex and more basic underlying entities.

Until recently, science has been dominated by analytic approaches; it is unlikely, however, that this dominance originates from their plausibility. For example, competing systems views have existed since at least the time of the Greeks (Bertalanffy 1968), and many individuals have considered these views to be more plausible than the analytic framework. There is little question that analytic approaches simplify the scientist's task, and it is possible that this accounts for their historical dominance. Analytic approaches did not offer a more compelling view of nature, but they alone often provided the mechanisms through which nature and methodology could be sufficiently simplified to permit scientific study and conceptual progress. The close parallel between the recent growth of systems approaches and the development of computer science does not seem merely fortuitous. In significant part, the computer's capacity to manage far more complex information than can the unaided mind appears to have given us the information-processing capacity needed to move forward in the development of systems approaches.[15]

Reinterpreting Common Occurrences

Certain common scientific occurrences may be partially explained by scientists' cognitive limitations, rather than by the other factors that have been proposed. Incommensurability and "groping towards maturity" are two possible examples.

INCOMMENSURABILITY

The problems involved in constructing neutral tests of different theories are widely recognized. According to Kuhn (1970), these problems are a product of the differing world views held by opposing theorists. Kuhn suggests that differing world views are associated with different habits of "seeing." His discussion of the processes or mechanisms that might underlie such habits, however, is limited to rather vague conjectures about perceptual learning. Judgment studies offer clues about what these underlying processes might be and raise challenges to Kuhn's explanation of incommensurability.

The judgment research provides strong evidence that individuals cannot adequately attend to all relevant information if the complexity of this information exceeds even minimal levels. Rather, they attend to information selectively, placing heavy emphasis on one or a few cues and underweighting all other cues. Further, individuals often lack insight into their internal judgment schemes. They overestimate the importance they attach to minor cues and underestimate their dependence on a few major cues, sometimes to an extreme degree.

These findings suggest that incommensurability can occur when researchers heavily weight and selectively attend to different cues. Failures in communication, which appear frequently to accompany incommensurability, can result when *reported* and *actual* policies for evaluating information conflict. Assume that two researchers select and weight cues or variables differently. One researcher may report that he or she attends to a range of variables, many of which overlap with the variables a second individual reports that he or she attends to; but these self-reports may diverge markedly from actual judgment policies. As a result, the two researchers cannot communicate, regardless of whether each can comprehend the other's report. Neither report is accurate, and neither researcher has access to either set of actual decision policies. Therefore, at best, the two researchers unintentionally mislead each other because their erroneous "self-insights" preclude accurate self-representation. Even if one refuses to accept the other's description of the reasoning underlying his or her conclusions, no conventional method is available for uncovering the actual decision policies of the other scientist.

As an example of the mechanisms by which different judgment policies can create incommensurability and communication failure, one can imagine the following situation. Jones holds a position that is largely founded on a set of studies he considers definitive. Smith, who holds a different position, agrees that these studies are definitive, agrees that they contradict her own position, and even agrees that the findings are of major importance. Assume, however, that we are granted access to Jones's and Smith's actual judgment schemes and find that while Jones does weight these studies heavily, Smith actually accords them little weight. Therefore, the studies have little influence on Smith's position. As a result, Smith may honestly report that she agrees completely with Jones's interpretation of the studies, but to Jones's amazement does not agree that these studies provide any *convincing* evidence against her own position. Smith may even be able to say why she is unconvinced, but her explanation is incomprehensible from Jones's point of view. Jones labels Smith as irrational, but neither researcher recognizes the true source of disagreement. What Jones weights heavily, Smith does not; and vice versa. Studies that are weighted differently cannot be neutral. Nevertheless, the two researchers cannot report or communicate the actual weighting policies they are using and may not realize that their policies differ. Therefore, each researcher's behavior may seem irrational to the other. Is it possible that the following quotation from Heisenberg (1973) represents such a case in which two scientists agree about experimental results but disagree about conclusions because of underlying differences in weighting schemes?

It was a very nice afternoon that I spent with Einstein but still when it came to the interpretation of quantum mechanics I could not convince him and he could not convince me. He always said, "Well, I agree that any experiment the results of which can be calculated by means of quantum mechanics will come out as you say, but still such a scheme cannot be a final description of Nature." (p. 392)

This explanation of incommensurability and communication failure contrasts with and potentially adds to Kuhn's explanation. For one, it provides a different psychological explanation for these phenomena, one based on cognitive reasoning strategies rather than the perceptual habits Kuhn describes. Second, it suggests that an additional process may contribute to the development of world views—that learning or developing a world view may be linked with learning to selectively attend to and heavily weight particular cues. One may gradually develop habitual practices for selecting and weighting cues that significantly determine how subsequent information is interpreted. Third, at least in some cases, differing judgment policies alone seem sufficient to create incommensurability and communication failure. Although the differing policies may be associated with broad and differing theoretical world views, they can also arise from less global differences in world view or from factors other than theoretical perspective. For example, minor differences in theoretical viewpoint or contrasting intellectual strengths and weaknesses may lead two scientists to employ divergent selection and weighting schemes. Broadly differing world views may be a sufficient condition to create incommensurability, but not a necessary one. Therefore, attempts to explain all cases of incommensurability on the basis of broad differences in world view may be misleading and add unneeded theoretical "baggage." Differing weighting policies may also help to explain occurrences in which, as described above, researchers agree that experimental results are accurate and can "see" experiments as others do, but still reach different conclusions.

Fourth, the cognitive perspective potentially allows one to explain cases in which commensurability and adequate communication exist, and it provides a methodology for testing this explanation. Specifically, one would hypothesize that the greater the similarity among the selection and weighting policies of different researchers, the greater the likelihood of valid comparisons and successful communication. To evaluate this hypothesis, one could use currently available judgment methodology to assess the selection and weighting schemes used by opposing scientists. The methodology could be applied to the judgments of contemporary and possibly even past scientists. One would then examine the relationship between the similarity of schemes and participation in "neutral" tests. This would require some independent method for identifying neutral tests, an obviously difficult but not impossible task.

GROPING TOWARDS MATURITY

Movement towards scientific maturity is usually tortuous. Slow, uneven, and laborious progress is far more typical than the leaps in knowledge that are so widely publicized. Scientific communities seemingly have extraordinary difficulty reaching consensus or arriving at impressive and reliable exemplars, two achievements associated with maturity. Masterman (1980) is certainly not reserved when describing such difficulties in making progress: "In real science, the thinking which goes on is enormously cruder, with the crudity going on much longer, than anyone who had not actually observed it going on would have believed possible" (p. 85). It is often assumed that progress is limited by inadequacies in the data base and that the rate of movement towards maturity is restricted by the rate at which scientists can generate more sufficient data. Perhaps, however, it is not so much limitations in the information that impede progress as limitations in human judgment. A less extreme hypothesis, but one that seems so obvious as to sound trivial, is that limitations in both information and cognition impede progress.

For example, although limitations in information might partially account for the difficulty reaching consensus, cognitive limitations also seem to be a significant factor. Because of scientists' cognitive limitations, their judgments may generally be of low validity. When low judgment validity is common among a group of individuals, they tend to produce widely varying opinions. Among groups of scientists, low judgment validity may not only create a diversity of opinion, but may also typically preclude any group member from constructing theories or explanatory networks that successfully integrate data and thus gain broad appeal. The obvious result would be difficulty in achieving consensus.

The achievement of consensus may be closely linked to the discovery of exemplars. This latter task itself is by no means simple, and the extensive efforts that are usually required may also partially stem from cognitive limitations. But once uncovered, exemplars probably simplify decision making. Because exemplars are impressive and highly salient demonstrations of puzzle-solving capacity, it is difficult *not* to attend to them, regardless of one's viewpoint. As the judgment literature shows, when highly salient cues are present they frequently are accorded heavy weight and other cues are underutilized (e.g., Tversky & Kahneman 1973). For many researchers, then, it is possible that exemplars become the limited set of cues that are weighted heavily and that largely shape the evaluation of information. Given cognitive limitations, other cues or findings are unlikely to have an equally powerful influence. Even if the exemplars arise within a framework containing obvious problems, their salience should ensure that the problems have only a secondary impact on judgment. Therefore, by providing salient cues that many researchers weight heavily, exemplars may significantly contribute to the achievement of consensus.

Cognitive limitations, then, may be a factor in the slow movement towards consensus, as well as in the generally slow movement towards scientific maturity. It would not be surprising to find, assuming we were clever enough to conduct such investigation, that the data available to past scientists often contained information sufficient for achieving far more advanced theoretical constructions than were actualized at that time and that cognitive limitations precluded such constructions. To an undetermined extent, limitations in human cognition, rather than limitations in information, restrict the rate of scientific progress.

RELEVANCE OF COGNITIVE LIMITATIONS TO PRESCRIPTIONS

If descriptions of science are altered by the study of cognitive limitations, what impact might this have on prescriptions for science? While it is popular to argue that description has no bearing on prescription, one can immediately challenge this proposition. To do so, I will discuss three basic approaches to the development of prescriptive programs for science and the interrelationship between description and prescription that characterizes each. The three approaches will initially be treated as if they were distinct categories, although in actual practice philosophers often combine them to varying degrees.

Descriptive changes directly and invariably force changes in prescriptive programs founded on what is now considered a naive view of science. These programs, which can be labeled "naturalistic," are based on the following assumptions: Science is a successful method for obtaining knowledge; the way scientists operate accounts for this success; and, therefore, the way scientists operate is a (the) prescribed method for acquiring knowledge. These assumptions are by no means completely unreasonable, but they are unreasonably extreme. In this framework, the actual practices of scientists comprise the one and final criterion for determining prescribed practices. Independent criteria, such as whether these practices are rational, and *critical* examination of the link between practice and outcome are secondary or irrelevant. Although few researchers fully embrace this prescriptive position, remnants still abound. For example, during meetings of the Society for the Social Study of Science, I have repeatedly heard individuals argue that a scientist's modus operandi might have seemed irrational or nonnormative, but that it still led to progress. This is taken as evidence that the scientist's particular practices should be prescribed. The tendency is to grant scientists unusual license to break rules of practice that have traditionally been considered rational or ideal and, when in doubt, to assume that a practice is desirable because it is part of science and science is successful.

Those holding the naturalistic position commit a form of the naturalistic fallacy. For our present purposes, there is no need to discuss this or the many other difficulties the naturalistic position creates (e.g., that one is limited to what has been or is and cannot move towards what could be). What is of interest is a logical consequence that follows from the naturalistic position. If this position is held to an extreme degree, then prescription must change when description of science changes, because prescription is directly derived from description. For example, if it is found that scientists do not use disconfirmatory strategies, then the use of disconfirmatory strategies is not a part of science and thus cannot be prescribed. Without exception, prescriptions found to be based on erroneous descriptions must change. For this reason, if the judgment findings were found to generalize to science and altered descriptions substantially, naturalistic positions would *have* to change.

A second type of prescriptive approach is usually tied closely to debates about the demarcation between "true" science and pseudoscience. These approaches may be partially founded on descriptions of scientific practices or outcomes. Unlike strictly naturalistic approaches, though, they are also based on considerations independent of "observations" or descriptions, such as principles of logic. The independence of these approaches from what is (science) provides their evaluative power and allows them to be used in attempts to separate true and pseudoscience. For example, Popper ([1935] 1959) considers refutability a necessary condition for true scientific practice. According to Popper ([1962] 1978), if a scientist constructs a theory that is not refutable, he or she is not truly practicing science. Although Popper's impressions of the actual manner in which scientists operate may have played a part in the formation of this criterion, it also maintains partial independence from any observational base. The criterion, then, represents more than just a summary of descriptions; it is founded on other considerations as well, such as principles of logic. Thereby, it can be used to evaluate whether an activity is scientific or not. Such criteria, whatever their source of origin, not only define true science but simultaneously serve as prescriptions for how one *should* practice science. For example: refutability = real science

= prescribed practice. One can label these programs prescriptive-descriptive. Lakatos (1970, 1972) and Popper are prominent philosophers who have adopted such approaches.

Given the independence of these programs from description, can description play a role in their evaluation and consequently lead to their modification? The answer is yes. Such programs generally contain implicit or explicit criteria for identifying "good" scientific activities that are partially independent of the procedures they prescribe. In other words, for something to be identified as good science, adherence to prescribed practices alone is not sufficient. Something else, usually some type of achievement or desired *outcome* (e.g., moving closer to truth or towards higher correspondence), also must be present. These programs claim that good outcomes can be accounted for, or are products of, adherence to the procedures they prescribe. Therefore, when episodes of good scientific outcome are identified, the procedures used by scientists should match with prescribed procedures. For example, if one prescribes refutability, one is simultaneously making the claim that good scientific outcome is associated with scientific practices that assure refutability. Therefore, if one identifies an episode of good scientific outcome but finds that the procedures or methods used by the scientist did not assure refutability, then this episode runs counter to the claims of the prescriptive-descriptive program.

Descriptions of science, then, can play the same role in prescriptive-descriptive programs that data do in evaluating scientific theories. One can even say that descriptions *are* the data. As with scientific theories, all data (descriptions) are not relevant; but descriptions that fall within the domain of the prescriptive-descriptive program (e.g., described episodes of good scientific outcome) are potentially relevant. Descriptions can serve as tests of the program, and inconsistent or disconfirming results are possible. In theory, either revisions of previously supportive descriptions or new descriptions that run counter to the claims of the program can provide disconfirming evidence.

Although in theory, the relationship between prescriptive-descriptive programs and inconsistent descriptive findings is straightforward, in practice the manner in which the two interact is far more complex and subtle. For example, one may insist that the prescriptions contained in these programs are the *sole* criteria for defining good or true science and that no other criteria are relevant. Whether this eliminates most of the history of science as truly scientific, as Feyerabend (1979) insists that Popper's demarcation criteria do, may be of little concern. As a second example, disconfirming descriptions may be dismissed for any variety of reasons, such as when one considers them inaccurate or inconclusive. The "data" of description, which are subject to endless indefinities and ambiguities, can be interpreted or used in any of a variety of ways because description serves the same role in prescriptive-descriptive programs as data do in scientific hypothesis testing. Definitive disconfirmation is not possible. But whether this allows one to claim that description is irrelevant is another question. One must account for the descriptions that fall within the domain of the prescriptive-descriptive program. If one who wishes to account for successful science claims (and prescribes) refutability as critical for success, then the repeated uncovering of successful scientific episodes in the absence of refutability places this claim at risk. That we know little about when prescriptive-descriptive programs should be altered in the face of inconsistent evidence, or about the processes that lead to their modification, simply reflects our ignorance of these matters in the arena of scientific hypothesis testing.

It has been argued that changes in description either should or can have an impact on naturalistic and prescriptive-descriptive programs. Is this also the case with programs that prescribe new methods of procedure that are hypothesized to lead to better science? The formulation of these programs, in principle, is not constrained in any way by actual practices. For example, given what is known about cognitive limitations, one may prescribe actuarial procedures for theory evaluation even though scientists do not now use these procedures. With such programs, are descriptions of science pertinent and what impact can they have?

Descriptions are hardly irrelevant to newly proposed procedures, although they do not necessarily hold logical claims. First, current descriptions of science often influence the construction of these prescriptive programs. Efforts to construct new procedures, for example, are frequently directed towards aspects of science that appear problematic, and the procedures are specifically designed to overcome perceived causes of failure. Description provides a context from which new prescriptions arise. Different descriptions are likely to channel prescriptive attempts in different directions and thus lead to divergent programs. Second, whatever the role of description in the construction of prescriptive programs, its role in the evaluation of the programs is critical. If one hypothesizes that the application of these procedures will improve science, then one needs to evaluate whether their application is possible and whether it will result in the desired outcome. Empirical evaluation may not be possible initially because programs can outline procedures or methods that must be developed over time. Further, some forms of evaluation may not require observation of outcome, such as evaluation conducted through thought experiments.

Ultimately, however, prescriptive programs that are intended for application must pass empirical tests of feasibility and outcome; one can only determine whether they do by observing (describing) attempts at, and the consequences of, their implementation. If it is found that new prescriptions are infeasible or do not lead to hypothesized outcomes, their claims are weakened and they are usually modified or abandoned. Therefore, even if prescriptive programs are not *initially* constrained by descriptions of science, they are often subsequently constrained or modified by descriptions of their feasibility and outcome—descriptions of how they work in actual practice.

Descriptions of science, then, are potentially relevant to any type of prescriptive program that is designed for application to science. This conclusion does not seem at all remarkable if one views these prescriptive programs in perspective. They are based, in part, on assumptions about the world, especially on assumptions about the functioning of human minds. For example, the logical positivists assumed that human logic is essentially a product of language or linguistic structures, and this assumption was central to their formulation of prescriptive programs. All prescriptive programs for science contain such explicit or implicit assumptions about the world, and in this sense they can be said to be theory laden.

It would be difficult, if not impossible, to construct a prescriptive program for science without a set of explicit or implicit assumptions about the human mind. This requirement at least holds for prescriptive programs intended for application, whether such application is to occur now or in the future when human knowledge and intellect are further advanced. Assumptions about mind have a substantial impact on prescriptive formulations, both in setting boundary conditions for determining present or future possibilities (e.g., whatever is judged to be impossible is not prescribed) and in designing the reasoning procedures that scientists are to use. Many of Bacon's prescriptions were designed to counteract tendencies towards bias. If Bacon did not believe human minds evidenced such tendencies, he would not or could not have addressed the problem.

The realities, or assumed realities, of human mind create the nasty practical problems with which one must dirty one's hands when designing prescriptive programs. One may of course attempt to ignore human realities and construct programs of abstract purity. If prescriptive programs are intended for application now or in the future, however, they must be constrained and designed in accordance with one's assumptions about the human mind as it is or could become. One does not, for example, prescribe divine inspiration as a means for testing hypotheses because we do not believe that we are capable of performing valid tests in this manner. To ignore practical constraints can result in programs that are of high intellectual and aesthetic merit but that alienate the scientist from the philosopher because they make, or even insist upon, impossible demands.

Given the dependence of prescriptions for science on assumptions about the world, it is not surprising that major shifts in description often eventuate in major changes in prescription. As discussed above on p. 36, revised descriptions of the world stemming from scientific study have resulted in changes in epistemology. This is not to say that description of science determines prescription for science (except with extreme naturalistic programs), nor to deny that various other considerations or procedures are relevant in the construction and evaluation of prescriptive programs. It is also not to argue that prescriptions should be altered when any change in description occurs. There are good reasons to exercise a conservative attitude towards revision. But it is apparent that immense gaps and shortcomings exist in our knowledge of what the scientific mind can do and how it operates. Thus, major changes in description are likely to be forthcoming, changes that will ultimately produce significant shifts in prescription.

To return to our central focus, the study of human judgment suggests that our current impressions, or descriptions, of scientists' cognitive capacities are seriously in error. For example, virtually all prescriptive programs requiring complex scientific judgments probably demand greater cognitive capacities than individual scientists or scientific communities possess. If in this and other cases we hold mistaken assumptions about cognitive capability, then, in accord with the above arguments, one can expect not only major changes in description but major changes in prescription as well.

REVISIONS IN PRESCRIPTIONS FOR SCIENCE

If prescriptive programs do require revision, what specific changes might be needed? The purpose of this section is to outline some possible approaches and directions for revision, particularly those that could aid scientists in overcoming the limitations that may exist in their capacity to carry out prescriptive dictates. There is no illusion that the possibilities raised below can be pursued with ease. Rather, one is faced with very involved technical problems, a series of trying conceptual issues, and many points of dispute. Nevertheless, the gains that may be realized seem to justify concentrated effort.

Prescriptive programs that require scientists to do what cannot be done under the conditions of current scientific practice could be revised in at least two basic ways. First, one could eliminate requirements for judgments that are beyond human cognitive capacities. Second, decision aids could be added to current programs without changing their basic structure. Decision aids would provide a means for overcoming the cognitive limitations that hinder the productive implementation of these programs. It is towards this latter possibility that I will direct the discussion.

Decision aids, as conceptualized here, encompass a broad range of activities and external mechanisms for improving judgment and, at times, transcending cognitive limitations. Any means for improving human judgment can be considered a decision aid, although the boundary between aided and unaided judgment is imprecise. The range of available aids becomes apparent if one recognizes that virtually all problem solving and decision making are multifaceted. These facets include such related activities as the collection, organization, and interpretation or evaluation of information. Decision aids can be applied to any one of these facets. The physician's use of X-rays can be considered an aid for the collection of information. A computer-assisted program for graphically displaying data is a decision aid for organizing information. A variety of aids are available for interpreting information, such as statistical tests, theoretical systems, and systems of logic. Some decision aids for interpreting information require only the implementation of internal thinking strategies, and one should not equate aids for interpretation, or other facets of decision making, with the use of external instrumentation or information-processing systems. Some aids are partially or fully internal. The scientific method provides an illustration. It encompasses all three aspects of thinking—the collection, organization, and evaluation of information—and provides decision aids that vary in the extent to which they are internally or externally based. For example, designing experiments, testing hypotheses, and using statistical tests or theories to evaluate information cover these three aspects and vary in the extent to which they are dependent on internal versus external aids.

How might one revise prescriptive programs to include decision aids that extend scientists' capacities to make various judgments? One possibility is to use actuarial methods to evaluate theories. We have seen that restrictions in the human capacity to perform complex judgments likely limit the effectiveness with which scientists can evaluate theories. Current methods of theory evaluation are essentially clinical (in the sense of the definitions of clinical and actuarial judgment provided on pp. 43-44). Given the consistently demonstrated superiority of actuarial methods, we may find that actuarial methods for evaluating theories substantially outperform currently used methods. The basic strategy for actuarial evaluation would involve the identification of signs or parameters associated with preferred theories. The preferred form would serve as the criterion for evaluating the theory, or theories, under consideration. A theory would be assessed for its standing on the identified signs or parameters, and, on the basis of these results, an actuarial evaluation of the theories approximation to the preferred form would be derived.

Objections have been raised to this proposal for actuarial theory evaluation. Some of these are based on misunderstandings and can be quickly dismissed. A common misunderstanding is the belief that the outcome of the actuarial evaluation alone determines the "fate" of the theory. This is not the case. Actuarial evaluation provides information about the relationship between signs or parameters and some criterion. The actuarial method offers more accurate assessment of these relationships than can be derived in one's head. How this information is used is up to the judge or scientist. (Whether it would be useful and how it could be applied is taken up in detail in chapter 5.) The actuarial evaluation does not magically determine the fate of the theory: it only provides a piece of information, which the scientist can use as he or she chooses. Another misunderstanding is the view that actuarial evaluation is a one-shot procedure that produces an immutable rating. But there is no reason why a theory cannot be continually reevaluated as more information is gathered or as the theory is altered.

Questions about the feasibility of deriving criteria for evaluating theories are worthy of more serious consideration because they raise deep issues about criteria for theory evaluation. Even if we have not or cannot establish criteria, however, actuarial theory evaluation is still possible. At least one actuarial procedure, that of "bootstrapping," can improve judgment even when no criteria have been identified. To bootstrap, one constructs programs that model the decisions of human judges. For example, one can abstract the judge's decision-making rules or procedures and use them to construct a formal decision-making program. If the criteria used by the judge are unclear, or even totally unknown, successful bootstrapping usually can still be accomplished. Bootstrapping only requires access to the information available to the judge and to the judge's decisions, and it does not require access to the criteria. To maximize the improvement in judgment gleaned through bootstrapping, one attempts to identify the more competent judges and to model their performance. The resultant product, or decision-making procedure, will not only outperform the less competent judges, it will usually also outperform the more competent judges, even if they are the very individuals upon whom the procedure was based. The reasons for this ascendance are unclear, but it may be a function of reliability — the procedure is not subject to such factors like fatigue or inconsistency, which lower the judge's performance relative to that which could be achieved given perfect reliability (Goldberg 1970). In any case, even if criteria for theory evaluation are not available, it still might be possible to bootstrap outstanding scientists and potentially achieve gains in judgment performance.

One could attempt to bootstrap either past or contemporary scientists as a means to investigate various issues. Assuming that we could use bootstrapping procedures to model the decision making of eminent scientists, analysis of the procedures might provide insights into their judgment strategies. Would analysis uncover systematic differences in the judgment strategies of eminent and noneminent scientists? Would it uncover judgment strategies common to outstanding scientists? And would bootstrapping programs for evaluating theories equal or surpass the performance of scientists?

Investigation of this last question poses interesting challenges, but there are many ways in which it could be attempted. Working from historical data, one might chart the success of a major theory over a period of time. The theory's success could be evaluated according to one or several criteria, such as the number of studies, puzzle solutions, or applied uses it generated. To assign a rating to the evaluations made of the theory, one could study the relationship between evaluations made at certain points in time and the success of the theory over subsequent periods. For example, one could analyze the relationship between an evaluation made at point A and the success of the theory over the period of time from point A to point B, the latter perhaps representing a time at which the theory was altered significantly. One could then look at the relationship between evaluations at point B and the success of the theory from point B to point C, the latter again representing a time at which significant changes were made.

Proceeding in this manner, one could obtain a series of evaluations and measurements of the theory's subsequent success, providing the data for analyzing the relationship between evaluation and success. Evaluations would be given high ratings if they corresponded with the success of the theory. A high rating would be obtained when a strongly positive evaluation was subsequently followed by a highly successful period. A low rating would be obtained when evaluations did not correspond with the success of the theory, such as when the theory received a positive evaluation and subsequently was not successful. Using this method, one could compare bootstrapping procedures with the human judge or scientist. One could rate prominent scientists' evaluations of a theory at different points in time and compare them with those obtained using evaluation procedures derived by bootstrapping these scientists. The method of judgment demonstrating a higher correspondence between evaluation and theory success would be considered the "winner," at least within the judgment domain studied.

Whatever the advantages bootstrapping might provide, it is still probably the least powerful actuarial method, and its use is best limited to situations in which criteria are unclear or unavailable. If criteria can be specified, more powerful actuarial approaches may be useful for theory evaluation. And we have just begun to explore actuarial approaches and other related decision aids that promise to greatly extend our cognitive abilities. For example, it appears likely that computer programs will be developed that can assess complex interrelationships, including those that might exist among criteria for theory evaluation (Findler 1979; Newell & Simon 1972; Simon 1981).

The argument for actuarial methods can be taken one step further into the realm of theory construction, which many would argue is of much greater practical relevance than theory evaluation. Cognitive constraints may not only limit scientists' ability to evaluate theories but also their ability to detect complex relationships among variables and, thereby, to construct theories. The rate of scientific progress is partially determined by the number of theoretical propositions and hypotheses that can be generated and the rate at which these constructions can be tested. Other things being equal, the more efficiently the scientist can conduct adequate tests of constructions, the greater the rate of progress. This rate will probably be further increased if one has a viable method for determining which constructions are most plausible and should be tested initially.

In this context, recent work in developing computer programs that can detect or form hypotheses about interrelationships among variables may eventually be of great importance to science in theory construction and evaluation. The proposal might sound whimsical, but there has already been one successful effort to pursue this aim. A computer program, given access to sets of data available to past scientists, was able to rediscover such things as Ohm's law, Galileo's law of falling bodies, and one of Kepler's laws of planetary motion (Langley 1979). In general, rapid advances have been made in developing the computer's capacity to detect interrelationships among variables (Baker et al. 1977; Findler 1979; Simon 1981). By all indications, this progress will continue and computers may soon far surpass human capability.[16] One can envision a time when computer programs will generate scientific hypotheses and theoretical constructions and evaluate them on the basis of available data and knowledge. For example, data pertaining to a scientific problem would be entered into the computer, which would then analyze them and generate a series of hypotheses or theoretical propositions. During or after this generation, such methods as actuarial procedures would be used to evaluate the constructions. Based on the outcome of this evaluation, the computer would list a set of hypotheses or theoretical propositions in order of decreasing likelihood (and perhaps even list additional needed tests). The findings might be used to help scientists determine what further experiments are indicated. The results of these experiments would then be entered into the computer, beginning the cycle anew.

Given what we know thus far about human cognitive limitations, the possibility (and at this time it is very much only a possibility) that such programs will bring about a leap in our capacity to generate and evaluate theories is far too important to ignore. Even in the face of our apparently limited cognitive powers, such as our difficulty in properly weighting more than a few variables or comprehending relatively simple configural relationships, scientists have accomplished some rather remarkable things. There is no telling what might be accomplished through major extension of our cognitive abilities. The use of computer-assisted actuarial methods for generating and evaluating theories, or perhaps other as yet unborn and unimagined decision aids, may bring about the next major step in the development of scientific method.

POSSIBLE REVISIONS IN METHODOLOGY FOR EVALUATING PRESCRIPTIVE PROGRAMS

The judgment literature suggests the need for revision not only at the level of scientific activity, but also at the level of metascientific study, or study in which science is the object of analysis. This need holds for those engaged in developing prescriptive programs for science in general and particularly for those developing methods to evaluate the programs. If new prescriptive programs are intended for application to science, one must evaluate whether their application is feasible and whether it results in the desired outcome. Philosophers often use clinical or intuitive methods to judge feasibility and outcome, and decision aids could likely be used to supplement and improve their judgments.

The scientific method, which can itself be considered a decision aid, can be applied to study the feasibility and outcome of prescriptive programs in science. We have seen that philosophical prescriptions for science include assumptions about means-ends relationships, about the ends that will result if certain means are implemented; the means are procedures or methodologies for conducting science, and the ends are some form of knowledge. The desired form of knowledge varies across philosophical systems, ranging from absolute truth to useful information (or some integration of the two), but all forms are ends nonetheless.

Certain decision aids, such as systems of logic, have been used in formulating and evaluating philosophical assumptions about the ends that will be achieved using specific methodologies, but we have rarely used the scientific method to evaluate assumptions about means-ends relationships. Some scientific methodologies can be used as decision aids for cross-validating means-ends assumptions. Philosophical programs could be reformulated as hypotheses: "If methodology A (or B or C) is used, the following outcome will be obtained. . . ." One could then test these hypotheses (see the research strategies described in the Appendix) by determining whether use of the specific strategy led to the predicted result. Scientific tests of the hypotheses could be conducted using the various decision aids associated with the scientific method, such as statistical analysis (Bayesian statistics and certain multivariate methods), actuarial analysis, and the objective rating of outcome. Use of the scientific method and associated techniques is a virtually untapped, potentially rich source for acquiring another window or knowledge base for evaluating these philosophical assumptions.

It is wrong to assume that scientific tests of philosophical assumptions about science are beset by impossible difficulties. Numerous assumptions are open to scientific testing, if we are only clever enough to develop the means and strategies for conducting investigations. An example towards which I have directed much attention is the scientist's capacity to perform certain of the judgment tasks required by prescriptive programs. Many of the judgments probably cannot be performed at the standard demanded by these programs. If further scientific testing supports this view, the results will provide strong grounds upon which to argue that components of prescriptive programs are not feasible, at least under the current conditions of scientific practice.

Using scientific "disconfirmation" of assumptions inherent in pre-scriptive programs to argue for revisions on the basis of pragmatic, if not logical, considerations cannot be considered yet another example of the vague, weakly supported statements about needed changes in science that have plagued sociological and psychological writings. Research on cognitive limitations has the potential to provide the social sciences with the first instance of a finding that convincingly challenges specific philosophical assumptions about prescriptive scientific procedures. Further scientific studies might provide additional instances.

CHAPTER 5

Reflections, Extensions, and Elaborations of the Argument

In this closing chapter, I will highlight and extend the central arguments already put forth and address some of the issues they raise. Many compromises are necessary in pursuing this aim, and I will often make only brief mention of topics that could easily provide material for an entire book. Topics were selected with the essential purpose of *this* book in mind — to establish credibility for the thesis that the judgment limitations of scientists are underrecognized and that these limitations have a *substantial* impact.

NORMAL OR REVOLUTIONARY PHILOSOPHY OF SCIENCE?

The major thesis depends on a series of underlying assumptions (see p. 3), which some readers may consider to be unreasonable; the entire argument for cognitive limitations would then be dismissed by branding the assumptions as radical and, by implication, invalid. I maintain, however, that none of the assumptions represents radical views and that all of them coincide with developments within the philosophy of science.

Let me start with two of the basic underlying assumptions, the first that scientific knowledge is a human cognitive construction and the second that the role of cognition in science should not be viewed as an undesirable contaminant but rather as an active, necessary, and unique contributor. When considering whether these assumptions are radical, one must determine against what standard they are to be compared. When compared with the views of science that dominated at the turn of the century, the argument that human cognition plays an indispensable role in the attainment of scientific knowledge could be considered pretty radical stuff. The assumptions might also be considered radical by hard-line empiricists holding the undying belief that absolute knowledge can be attained. But when the assumptions are considered in relationship to contemporary developments within the philosophy of science, they do not seem radical at all.

Many philosophers have gradually abandoned the quest for absolute objectivity and knowledge, a move that could be considered the major conceptual shift within the philosophy of science over the past thirty to fifty years. It is increasingly recognized that scientists have not, and cannot, reach either an incorrigible factual base upon which to build theories or, in part because of this, certain theories or certain ways to test them. As a result, logic's promise to produce scientific truth has been undercut: if there are no definite givens upon which to perform logical operations, then the empiricist plan for attaining truth cannot succeed. We cannot be sure about the factors that have lead to this conceptual shift and that continue to promote it. Criticism and discoveries within empiricist circles, external challenges and counterproposals from other schools of thought, historical studies of science, and developments within science (e.g., the uncertainty principle) have surely had an impact. Along with the doubts raised about programs for establishing absolute knowledge, many responses and reactions have been put forth, including attempts to save the programs, to avoid the tough issues completely, and to develop alternative programs. The proliferation and diversity of views make it increasingly hard to identify the "mainstream" within the field, but it is safe to say that many philosophers of science accept the impossibility of achieving pure objectivity or absolute knowledge and are now pursuing other alternatives.

The emerging realization that the search for absolute truth is an impossible quest has been linked with a still developing awareness that scientific knowledge is not automatically produced by the correct application of prescribed procedures; rather, it depends upon human judgment as a mediating link in the interpretation and construction of data. This awareness is reflected in many ways, such as in the "sacrifice" of scientific discovery to the cognitive domain (as opposed to the domain of logic); the identification of persuasion as a primary means for swaying scientific belief; and the recognition that scientific communities do not test single theories for their correspondence to nature, but rather compare a series of theories (or even paradigms) with each other. These developments within the study of science converge towards the identification of human cognition as fundamental to all scientific activities, although those studying science often do not explicitly realize this connection. But there is broad recognition that scientific activities and progress are dependent on judgment and not on a mechanical application of method in which the scientist strives to eliminate contaminating human influences. Some individuals have used arguments for the judgment-laden nature of science to found programs that deviate broadly from the mainstream. I do not think, however, that arguments for science as judgment laden and for the necessary role played by human judgment can, by themselves, any longer be dismissed as radical.

Are Arguments for the Role of Human Judgment Antiscientific?

Those who argue that human cognition is essential to science risk being falsely stereotyped as antiscientific. Therefore, I would like to go on record as stating that it is possible to view human judgment as essential to science without holding (and I do not) any of the beliefs listed below.

Belief 1: Science is irrational. Logical and rational judgment should not be conflated. Many scientific questions cannot be resolved through logic alone. Perhaps one cannot simultaneously be illogical and rational, but one can be rational without exercising explicit rules of logic. The growing awareness of science as judgment laden will force us to consider a new set of problems about scientific decision making that cannot be settled by deferring to principles of logic alone, such as when it is rational to retain a theory in the face of counterevidence. These problems might offend those who wish to limit analysis of science to the domain of logic. But one can consider such problems without believing that science is irrational. One can even believe that scientists have judgment limitations *and* that many scientific judgments *are* rational. And one can hold both of these beliefs and still maintain that scientific judgment could be substantially improved. There is more to the study of scientific reasoning than determining whether a judgment is or is not logical, and no one-to-one correspondence exists between rules of logical judgment that have been set forth for scientists and rational scientific judgments.

Belief 2: If logic doesn't rule, anything goes. The argument that scientific decisions depend on human judgment and cannot be settled by any final authority *is not* the same as the argument that any judgment is as good as any other. One can acknowledge the role of human judgment in science and still assume that there are better and worse judgments. The problem within science is that we often do not know how to distinguish good judgment practices from bad ones. Nevertheless, even if we cannot definitively distinguish better and worse judgments, "good judgment" can serve as a regulative ideal. One will not be so concerned about a lack of definite guidelines if one realizes that regulative ideals are not static but dynamic and develop over time. Whether some forms of judgment can once and forever be labeled as "bad" depends upon whether regulative ideals for science have a "fact-correcting" nature, a complicated issue that will not be addressed here. The point is that proposed definitions of good judgment, even in the absence of final answers, can serve as useful guides. One should not confuse our lack of final definitions with the merit of differential valuation of judgments, nor take from this assertion that current definitions, because they are not final, are of no use.

In many places in the book, such labels as "bad judgment habits" or "erroneous judgments" appear, labels that are often not difficult to justify when one is referring to formal judgment experiments. The experimenter frequently presents an artificially constructed problem with a set answer, a problem on which an error is an error. As Weimer (1977) points out, however, actual scientific practice is an entirely different matter. We have no absolute truths or answers, nor have we any sure ways to identify truth if we happen to hit upon it. In this book, when evaluative labels are applied to scientific judgments, they generally rest on basic principles of logic. One example is the use of a diagnostic sign that decreases the frequency of correct identifications relative to the level that would be achieved using base rates alone. It seems a simple matter to say that a scientist who proceeds in this manner, and believes that this strategy increases his or her hit rate, has made a judgment error. But even when there are compelling reasons to label scientific judgments as erroneous, the labels can only be considered hypotheses; we actually have very little explicit knowledge about good scientific judgment practices. One hopes that we will progress in our ability to recognize and detect "bad judgment habits" and judgment "errors," but perhaps we will never be certain about the labels, even when the merits of judgment practices seem self-evident. Again, one must be careful about conflating logic and rationality.

If the reader wishes to quibble with my use of labels like "erroneous judgment," based on the argument that all scientific judgments are erroneous in the absolute sense (i.e., are not absolutely true), he or she may substitute other terms, such as better and worse judgment, or progressive versus regressive judgment. But I do not endorse the argument that the topic of good versus bad judgment cannot be addressed at all or that all judgments should be considered equally acceptable, because we cannot identify correct scientific judgments in any absolute sense. I have no qualms about attempting to apply labels identifying scientific judgments as correct or incorrect, although I always question whether I or others have applied them appropriately.

Belief 3: Science is denigrated. Maintaining the belief that science is judgment laden is by no means incompatible with maintaining respect and admiration for science as a major human achievement. The former view has been used by some adherents as a springboard for various antiscientific platforms, such as arguments that scientific knowledge is nothing but a base reflection of class status; but there is no reason why one must draw negative implications. An awareness of the important role played by human judgment can lead to an increased respect for scientific activities. For example, if one believes that scientific data and method alone ensure the gradual attainment of truth, there is limited room for individual achievement. If we instead believe that the scientist as human judge "makes" scientific knowledge, then scientists, rather than the method, deserve the major credit for doing so. Individuals who fervently maintain that science is the means to final and absolute truth must see as blasphemous statements about the role of human judgment and the related impossibility of achieving absolutes. But beyond showing us that final truths cannot be reached, belief in the importance of human judgment does not necessarily lessen the achievements of science, or scientists, in any way. This is *not* an antiscientific book, and I have not intended for any of my statements to be antiscientific.

If my previous statements and arguments appear antiscientific, perhaps a good measure of the blame falls on me. The developmental perspective that I endorse bears a surface resemblance to antiscientific viewpoints, and I have not explicitly identified my developmental roots.[1] I view the growth of human knowledge, and methods for obtaining knowledge, from a developmental perspective. Science can be seen as one achievement or landmark in a long progressive path. Science may be our most advanced achievement, but I do not see it (at least in its current form) as exhausting the possibilities for growth in either knowledge *or method*. To view current scientific method as the end point, that is, as a final and perfect product, mimics the shortsightedness of those who believed, before the advent of science, that their methods for acquiring knowledge could be advanced no further. From the developmental perspective, even scientific knowledge and the methods for obtaining it are not final achievements but things that can advance further. Whether the advance will involve an internal expansion and growth of scientific methodology or the eventual development of an entirely new system for obtaining knowledge is a question not to be resolved here. Regardless, one can view science as *developing*, as a method that will go through further evolution; and one can applaud current accomplishments while believing that further developments can and should be made. I can see nothing antiscientific in maintaining these beliefs.

Two other assumptions appear at the beginning of the book. One is that scientific results do and should contribute to the philosophy of science; this assumption is not really relevant to the current discussion, and I will not explore it here. The second is that scientists are faced with difficult and complex judgment tasks. As I argued in chapter 1, this difficulty and complexity are present across virtually all scientific activities and at all levels of inference, even the factual level.

Are arguments for complexity at the factual level overdone, as one of my reviewers claims? If anything, I believe the case for complexity has been understated. Anyone who wishes to glimpse the complexity of processes needed to manage the sensory world might find current studies of the human brain illuminating. Eccles (1977), a prominent researcher in this area, makes the following statement when discussing human perception:

Before the cerebral cortex is involved in the necessary complex patterned reaction to the sensory input so that it gives a conscious perception, there will be activity of this immense, unimaginable

complexity. . . . it will be necessary to develop new forms
of mathematics, as yet unimagined, in order to cope with such
immense patterned complexities. (p. 204)[2]

One should not assume that a one-to-one correspondence exists between the complexity of internal human processes and the complexity of the external world, but Eccles's statement gives some grasp of the
incredibly involved internal processes that underlie the attainment of
meaningful perceptions. Whatever the exact level of complexity present in the sensory world or in factual perceiving, I take no risk of
doing violence to the word *complex* when I use it to describe various
facets of science and scientific judgment tasks.

Two additional assumptions, which were not mentioned directly
in earlier pages, underlie many of my arguments. The first is that scientists are not fully capable, or are less capable than is typically assumed, of managing the complex judgment tasks they face. Some of
the reasons why we may have formed mistaken assumptions about
scientific judgment prowess were covered in chapter 1, including the
potentially misleading nature of introspection and the tendency to
overestimate the typical complexity of successful scientific solutions.
These reasons may, however, be pseudosophisticated explanations.
The creation of false impressions may actually occur in an extremely
rudimentary fashion: our recognition that scientists have achieved
some rather remarkable things may lead us to conclude that they must
be "darn smart," and part of being "darn smart" is the capacity to
think in highly complex ways.

I would argue that scientists' cognitive limitations are widely underrecognized and that this stems from a fundamental attribution error. What scientists have not accomplished or solved is usually attributed to something other than cognitive limitations, whereas cognitive
limitations may be the most parsimonious explanation. For example,
scientific failures are often explained by such things as lack of sufficient data, a muddling of judgment by subjective factors or biases,
or the nature of the questions that have been posed (e.g., that they
are not appropriate scientific questions). Time and again, these are
the factors that are used to account for the upper boundaries of scientific knowledge and for unsatisfactory progress on scientific problems. Although all of these factors may operate, the most fundamental one may be our insufficient cognitive ability to answer questions
—or at least to reach satisfactory solutions at a satisfactory rate. This

other explanation for the limits of science was perhaps best summed up by Einstein when he was asked why greater progress had not been made on certain physics problems. He replied that the problems were "too hard for the physicists."

The possible boundaries that cognitive limitations place on scientific growth will be discussed below. My present aim is to raise the hypothesis that an important, if not primary, source is missed when the boundaries of science are attributed to factors other than cognitive limitations. At minimum, there is little doubt that we would progress more rapidly if we were smarter. It would take great daring to argue that the scientific method and human intellect have been perfected and that restrictions in scientists' judgment capacities are irrelevant or nonexistent.

Finally, to accept arguments for the relevance of cognitive limitations, one must accept that the current imperfections of science do not stem from a willful refusal to follow prescriptive dictates. One must believe that all of the problems of the scientist would *not* be solved, even if scientists made a true commitment to follow prescribed procedure. There is little choice but to concede this matter: aside from the question of whether scientists have the capacity to meet the judgment standards set by most prescriptive programs, it is clear that these programs leave many problems unsolved. Frequent deviations by scientists from these programs may not so much reflect stubborn refusal as an accurate appraisal of how useful they are. Further, we know almost nothing about what produces good science, or (at least at this point) we cannot translate our knowledge into language terms. In the face of our meager knowledge of these matters, it is naive to assume that adherence to prescriptive dictates would lead to ideal outcome. Strict adherence to current prescriptive programs would not bring such immense success that we could abandon discussion of factors restricting scientific growth.

Conclusions

I hope I have established the unremarkable nature of the claims underlying the major thesis. Despite argument about degree or extent, such as about the level of complexity present at the factual level, most readers will agree that there are tenable grounds to assume that science is judgment laden, that scientists are faced with complex and difficult judgment tasks, and that scientists are not perfectly equipped to manage the judgment tasks they face. An emerging awareness of these "realities" of scientific life, and basic developments within the philosophy of science and cognitive psychology, all point towards the same conclusion—that the study of scientific judgment and its limitations is essential to an understanding of the processes involved in the acquisition of scientific knowledge. This growing recognition will lead those in the field to restructure many longstanding problems and to pose a new set of questions. We must, for example, develop methods for identifying rational scientific decisions that are not founded on rules of logic alone, such as accurate decisions that a change in theory will best promote scientific aims. Many of these problems relate to issues that are often of greatest interest to the scientist—those of pragmatic action—but they extend into other realms as well. Examples include belief maintenance and change, capacity to integrate increasing amounts of information, and optimal judgment strategies. The study of these problems will certainly provide important lessons for the cognitive psychologist. Most of all, one hopes that the work will lead to a better understanding of fundamental issues in epistemology and, thus, to further refinement and development of scientific method.

USEFULNESS OF ACTUARIAL METHODS AND OTHER DECISION AIDS

Although my major thesis and its related assumptions may not arouse incredulous reactions among many readers, these reactions are typical when one proposes the actuarial evaluation of theories. My previous arguments for actuarial theory evaluation were only meant to provide one example of how scientists might use decision aids. The principal argument is that the further development and utilization of decision aids would be of benefit to scientists, particularly for "higher-level" judgments.

What Decision Aids Cannot Do

The goal of decision aids is *not* to provide a method of instant assessment or (stated in another way) a final authority for evaluating scientific proposals. This goal is not feasible because it requires final knowledge. For example, one could not conduct final evaluations without absolutely knowing the characteristics of true scientific proposals or theories. There will always be uncertainty about the criteria that are developed, as well as about the validity of methods used to assess the standing of theories along these criteria. This uncertainty remains whether evaluation entails the comparison of a theory to nature, alternative theories to each other, or paradigms to each other. Therefore, even if one does not believe in Kuhn's (1970) notion of incommensurability, the obstacle is in no way circumvented.

Proposals for the use of decision aids in science, at least as put forth here, are not a search for the Holy Grail. Decision aids are not a means for obtaining totally objective, theory-neutral evaluations. They may lead to improved methods for evaluating scientific ideas, or they may be applied beneficially in other ways; but they offer no promise for the attainment of certain knowledge.

What Decision Aids Might Do

Context of Evaluation

The first use that I envision for decision aids falls roughly within the context of evaluation, although I am referring here to evaluation as it is used in everyday scientific practice, not to the philosophical analysis of formal theories. Scientists do not limit evaluation to polished products. Evaluation, or critical examination, is a continuous process that takes place across all facets of science. Again, rather than critical appraisal of theories as a whole, I am referring to the scientist who asks, Is my notion about this occurrence correct? Is my equipment detecting what I think it is detecting? Is this the right experimental procedure for the question I am asking? Many of these evaluations occur internally, and only a small minority can be or are subjected to formal scientific test. Scientific effort is not just a completely blind and random search because the constant evaluation of thoughts and procedures occurs on this "mundane" level. Without evaluation there is no selectivity, and without selectivity science would be nothing but a fortuitous exercise.

Evaluation is impossible without some form of feedback about one's ideas or constructs. One can obtain feedback, or evaluative information, in many ways and on many levels. This process can be purely internal, as when one conducts thought experiments, or it may depend upon information obtained through scientific experimentation. What is critical for progress, or the development of thought, is feedback that adds to or alters previously held beliefs. If all feedback is simply incorporated into what is already known or believed, no change in thought will take place.[3] In science, thought-altering feedback often takes the form of information about the world that does not fit precisely within current beliefs.

All feedback is not of equal value. Some forms mislead the scientist, as when measurement artifact is confused with signal. Feedback that is lacking in clarity or refinement may also be of limited use. Developing the capacity to obtain or render feedback with increased precision is one of the primary means for enhancing its potential usefulness. The word *potential* here indicates that the value of feedback is ultimately dependent upon the scientist's capacity to glean information from what has been provided. When attempting to improve feedback, one may go beyond the level of precision that the subject matter permits or may lack the knowledge structures that are needed to benefit from increased precision.

If evaluative feedback is to be useful to the scientist, it must consist of something more than global nay-saying or yea-saying. Evaluations that inform the scientist that his or her theory as a whole contains some type of flaw are of little benefit. Virtually no scientist is so naive as to assume that the theory totally captures truth, and being told that this is indeed the case is of no pragmatic use. For this reason, Popper's early contention that theories should be abandoned in the face of disconfirming results offered no useful guide for pragmatic action and sat far removed from the realities of scientific life (see Lakatos [1968] for a review of early and later Popperian views).

Feedback can be of use to the scientist only if it offers more specific information than is provided by global evaluations of entire theories. For example, when standards or expectations can be developed for the outcome of specific experiments, feedback might indicate whether results did or did not fit within expectations. When results did not, repeated manipulations in procedure might be attempted, with continuous evaluation of the relationship between each attempt and its results providing useful guides. This process might occur on the level of formal scientific experiment, such as when one repeatedly attempts to create alloys that exhibit a certain set of properties; or at higher levels, such as when one forms a general postulate and reviews a set of studies to determine whether their outcomes fit within expectations. These forms of feedback can help inform the scientist about the soundness of his or her ideas, indicate when modifications in procedure or thought might prove productive, and stimulate the formation of new ideas.

As a field advances and scientists are able to ask more refined questions, more refined or precise feedback may be needed to address these questions. When expectations can be expressed in exacting mathematical terms (e.g., the results of this experiment should fall within this narrow range of numbers), sufficiently precise measurement techniques are required to determine whether these expectations have been met. Greater precision is frequently attained by developing more sensitive and specific measurement techniques, by quantifying observations, and by applying or creating mathematical methods for analyzing data; these attainments often occur in an interdependent manner.

Decision aids potentially offer the scientist a more systematic and precise method for obtaining evaluative feedback about scientific proposals. The level at which feedback is needed and the precision that is required depend on the maturity of the field and the types of questions that are posed; the latter often vary greatly in level of refinement both across and within fields. Within psychology, even very modest gains in the precision with which one can render data bearing on specific hypotheses may be valuable. Subjects in the Chapman and Chapman (1967) experiment on prior hypotheses and human figure drawings (see pp. 65-67) might have reached more accurate conclusions if they had simply kept a paper-and-pencil count of confirming versus disconfirming instances; clinical psychologists employing figure drawings might also benefit from this simple procedure. Many decision aids (such as statistical tests of significance) are available at the level of single experiments, but very few aids that incorporate such things as mathematical methods or computer technology are available for the higher-level integrative acts needed to evaluate a series of studies. In the social sciences, for instance, reviews of related studies often do not go beyond "counting noses" (i.e., five studies in favor and six against). Researchers might benefit from the development of formal methods and aids for evaluating sets of studies. At a higher level of integration, methods might be developed for evaluating the characteristics of theories; one might develop metrics for evaluating structural features of theories, such as their parsimony. One might also try to measure the complexity of the reasoning steps by which scientists account for findings that deviate from initial theoretical expectations. Complexity that exceeded a certain level might serve as a "warning signal."

In summary, decision aids for evaluation might enhance the usefulness of feedback by increasing the precision with which it can be obtained. Possible developments include mathematical methods and means for conducting higher-level evaluation. Evaluative procedures that produce numerical results can in turn permit more precise evaluation of the procedures themselves, thereby facilitating their further development. For example, one can apply mathematical techniques to study the relationship between evaluation outcome and other variables, such as progress within a field. This use of decision aids does not approach the goals of those who have labored unsuccessfully to develop methods for the instant evaluation of a theory's truth status, but perhaps these goals have precluded success and the development of programs that are useful to the scientist.

Context of Discovery

Decision aids may also help scientists to discover new things about nature. If one views decision aids broadly as ways to increase our capacity to collect, organize, or evaluate information, it is clear that they have already contributed to scientific discovery. Specific aids— diagnostic techniques in neuroscience, such as computerized axial tomography—and general ones—mathematics—have contributed to scientific progress. At issue, then, is not whether decision aids can be useful, but whether their use can be productively extended into additional realms of scientific activity.

Our particular interest is the possibility of developing methods for generating scientific hypotheses and perhaps even higher-level constructions, such as laws or theories. We can identify at least two basic issues related to this possibility. The first is whether the information provided by decision aids merely reflects what is already known or whether it can add to existing knowledge. In other words, are there times when the information serves as the leading edge in the development of knowledge? Consider this question relative to the role of mathematics in science: Scientists cannot, in the Baconian sense, collect mathematical "observations" and build a scientific knowledge base directly on them. If building a scientific knowledge base were so straightforward, the Baconian sciences would not have taken so long to develop. Rather, the abstraction of scientific meaning from numbers requires some preliminary conceptual development. Without some basic ideas or concepts, there is nothing to guide the collection and interpretation of information. Representing, organizing, or manipulating observations and ideas in the form of mathematics, however, can spur new conceptual developments or suggest new hypotheses and ideas—mathematical products can take a leading role in the development of knowledge. If decision aids like mathematics can perform this function, if mathematical representation provides more than just a shadow of what is already known or makes a unique contribution, then other decision aids may perform a similar function. Therefore, if one can develop decision aids that construct hypotheses or even theories, these aids may contribute something unique to the development of scientific knowledge.

The second basic issue can be stated in question form. Is it possible to develop external mechanisms or machinery—methods that stand apart from the scientist—that can generate hypotheses or theories? To what extent would (could) such constructions go beyond what is already known? Although it is impressive that a computer program was able to rediscover scientific laws (Langley 1979), it would have been more impressive if the computer had done so before the scientists—and more impressive still if it had been accomplished before all of the data with which these scientists worked had become available.

The first of these two issues can apparently be resolved without great difficulty. In theory, decision aids can make a unique contribution to scientific discovery. The second issue resists easy resolution. The possibility of creating external means for generating hypotheses or theories that *go beyond* what is already known now appears to depend upon developments within computer science. Computers have already contributed to scientific discovery by organizing or analyzing data in ways that provide suggestive leads, but we do not know whether they can step one level beyond this achievement and directly discover new things. This step beyond might, as one possibility, entail the capacity to select from among humanly generated alternatives the one hypothesis that best fits a set of data or the capacity to uncover new instances of application (or fit) for previous hypotheses—to extend them successfully to new domains. As a second possibility, the step beyond might entail the capacity to generate original hypotheses or theoretical structures. Roughly speaking, the difference in Kuhnian terms between the first possibility and the second one is the difference between extending normal science and creating revolutionary science, respectively.[4]

Although we can imagine, at least in theory, how the first possibility can be accomplished, the second one seems to defy reason. How can a computer generate original ideas or knowledge structures that are not stored within it in the first place? This requires the capacity to "learn," that is, to go beyond the original programming and acquire new concepts. This capacity to, as it were, pull oneself up by one's bootstraps might seem impossible to achieve, but it is commonplace in human development. It is self-evident that human conceptual progress could not have occurred had we not been able to go beyond what we already knew. Further, Piaget's work (1970) provides extensive documentation that cognitive structures undergo progressive reorganization during the course of development. Individual cognitive development is not simply a matter of adding new facts. It also involves the reorganization of thought into successively more advanced structures that allow us to perform previously impossible reasoning operations and to better comprehend the world. These new structures are not attained through direct learning, that is, they are not taught directly to us by others. Although experience is necessary for their attainment, experience elicits and generates internal thinking processes that eventuate in internal reorganization. These new and more advanced structures are not given to us but are gained by us, and in this sense we pull ourselves up to new conceptual capacities by our bootstraps. With apologies to those already well versed in Piaget, I raise this point about human development to show that new and more advanced concepts can be gained through internal processes and need not be directly furnished by external sources. Obviously, there are means and mechanisms through which this occurs; but our lack of knowledge about them can make the process seem surrealistic.

If individuals can pull themselves up to new concepts by their bootstraps, there is no reason in principle why computers cannot do the same thing. Investigators have successfully programmed computers to learn from experience. Samuel's (1967) creation of a chess-playing computer that learns from mistakes provides an example. We do not know, though, whether we can create computer programs that not only learn new applications of things they already have been told (know), but that generate original "concepts." Developments within the computer sciences offer promise for the creation of such capacities (for overviews of these developments and supportive positions, see Arden [1980], McCorduck [1979], and Simon [1977, 1981]; for an opposing view see Weizenbaum [1976]). But a determination will require the test of time. Entirely new forms of technology and machinery may also be developed for generating ideas.

If through the computer or some other means we can develop the capacity to generate hypotheses or theories externally, would the information be useful and how would it be used? To what extent would we be able to grasp and utilize what has been generated? Will we see the era of the computer as sage? When we pose technical problems to this sage (e.g., how to treat certain diseases), will it directly instruct us how to proceed? When we wish to understand certain phenomena better, will it have to teach us what it has learned and will it know things that, given our cognitive limitations, it cannot teach us?

These questions may strain our images of the possible, but decision aids have many potential uses that are far less "wild." For example, physicians struggling to identify their patients' ailments are already benefiting from centralized data banks and computer programs that aid in diagnosis. How far we will progress beyond the technology and methods that have been thus far developed for assisting human judgment, and how far such methods can go in extending what we already know, are questions that we cannot even begin to answer now. They are questions, however, that will surely evoke a great deal of interest and concentrated effort over the coming years, and perhaps developments that may bring major changes in our world.

SIMPLE MODELS – SIMPLE WORLD?

Those who have read this far probably believe that questions about the cognitive limitations of scientists pose important problems. Throughout the text, I have discussed these problems from various angles and have drawn various implications. But I have not directly addressed what is perhaps the most fundamental issue raised by the judgment literature.

To the question, What poses the greatest obstacle to the progress of science? I would answer, human cognitive limitations. To the question, What aspect of human cognitive limitations poses the greatest obstacle? I would answer, restrictions in the capacity to reason in a complex configural fashion and thus to grasp complex configural relationships. The most fundamental question, though, is this: What are the boundaries in our capacity to know the world, given our apparently limited capacity to reason in a complex configural manner? In other words, to what extent will the secrets of the world yield to the reasoning capacities we possess or can employ?

One possibility that might render the above issue a vacant one is that only simple principles, and thereby only simple reasoning, are necessary to explain the world.[5] To "prove" this proposition, one would have to explain the world adequately with simple principles. Obviously, the proof cannot be expected to be immediately, if ever, forthcoming. Inability to obtain it does not prove the opposing proposition that the world is complex, for it can still be maintained that the world is simple and we have just not been able to find the right keys. There will continue to be legitimate grounds for maintaining both the proposition for simplicity and that for complexity. But at least for those who view the world as complex, the question of how well this world can be understood or represented in the face of our cognitive limitations will probably be considered a relevant concern.

The importance of the problem is continually dismissed or grossly underrecognized. There is a general faith that scientists will be able to solve the complex problems with which they are faced if they can only obtain adequate data, and examples that run counter to this faith are freely ignored or given minimal weight. The immense struggles of the "soft" sciences, for example, are often easily explained away as products of their immaturity. An alternative explanation is that successful scientific fields have achieved this status because, unlike the soft sciences, they have been able to locate problems that are simple enough to solve. Scientific successes, then, may not so much represent the solution of complex problems as the solution of problems that yield to and "tolerate" simplified ways of viewing the world. Holton (1978) argues that many of the early and impressive accomplishments of astronomers were possible because these scientists had the good fortune to uncover problems that could be satisfactorily resolved by simple thinking.

The . . . data had to yield a simple system eventually, for the plants *were* independent of one another. . . . The planets formed therefore a "pure" sample, possibly the simplest of all the samples with which science has had to deal ever since, and hence the success of early systems makers of astronomy was a success of a philosophy of oversimplification. . . . (p. 142)

In the soft sciences, such as psychology, it may be exceedingly difficult to find problems that yield to simplification. Meehl (1978) lists twenty features that complicate study in psychology. Included are the immense difficulties dividing the "raw behavioral flux into meaningful intervals," the large number of variables that can and do have an impact upon human behavior, and the problem of uncovering critical events that influence personality development. Regarding the latter, Meehl uses as an example "inner events, such as fantasies, resolutions, shifts in cognitive structure, that the patient may or may not report and that he or she may later be unable to recall" (p. 810). Scientific progress, then, may only be possible to the extent that problems permit simplification. Further, some types of problems may be very resistant to simplification, and limits in our capacity to think complexly may set strict limits on our capacity to grasp the world. Feyerabend (1979) recognizes this problem when he says that "we are also undercover agents in nature, always trying to adapt it to our simplistic conceptions and never succeeding" (p. 92).

As I hope is clear, I do not view our limitations in the capacity to perform complex judgment as an abstract or distant barrier that we may someday face. Rather, such limitations always have, currently do, and will continue to restrict the success of virtually all scientific effort. If we were capable of greater cognitive complexity, we no doubt would have made far greater scientific progress than we have achieved. The question then is how far we can progress in the face of our cognitive limitations. This question goes beyond concerns raised about limits in our capacity to obtain revealing information about the world. It is more basic. One must ask, if nature were somehow revealed to us, how much would we be able to grasp of what we were shown? Our ability to transcend our current cognitive limitations, either through decision aids or some other means, may ultimately determine how much we will learn about the world and fulfill scientific goals. The attainment of means for transcending cognitive limitations perhaps has been, and will continue to be, the primary mover in intellectual development.

I would like to finish this work on a personal note. I have found the study of human judgment a profoundly humbling experience. The recognition that many beliefs about the power of human judgment are probably more fiction than reality can lead to despair, but I do not think that negative reactions need to dominate. As Gandhi tells us in the quotation placed at the beginning of this text, the recognition of limitations can ultimately provide a source of strength. An increasing awareness of cognitive limitations in science, by clearly defining and outlining a problem, can lead to productive effort. One can be aware of our cognitive limitations without abandoning hope that we can move beyond them. In achieving an awareness of limitations, one gives up something, something about themselves and about others. But this loss may be the exchange of a false idol for greater enlightenment.

It is my belief, and I freely admit that it rests on faith, that through recognition and effort we will be able to extend our reasoning capacities. What is to be gained by such an achievement is knowledge for its own sake and a greater capacity to predict and control the events around us. When considering our capacity to use such powers in a consistently just and humane fashion, however, my faith breaks down and I must acknowledge grave concerns. Extending our cognitive capacities may make us smarter, but it will not necessarily make us wiser or bring us closer to ethical perfection. It is my hope that, in pursuing the apple, the latter goal will not be sacrificed.

Appendix

Research Strategies

How might one go about studying some of the issues that I have raised about scientific judgment? This Appendix provides a survey of four basic research strategies, which are not necessarily exclusive of one another and can be combined. The strategies offer an organizing framework for viewing potential approaches to judgment research, and the discussion may also serve to stimulate and organize the reader's own thoughts about the study of scientific judgment.

HISTORICAL STUDIES

Historical study provides unique access to a valuable data base. Besides supplying additional instances of phenomena germane to areas of interest (great discoveries and scientists, successful and unsuccessful theories), past occurrences also offer a means for studying phenomena over time. The development and subsequent utilization of a major scientific invention is one example. Although historians of science have made distinguished contributions to our knowledge, they have made little use of more sophisticated methods for combining and analyzing data, such as multivariate statistical techniques. There is no reason why these techniques cannot be used in the historical study of scientific judgment, and it is towards this possibility that the following examples will be directed.

First, by capitalizing on the enlarged pool of instances of particular phenomena made possible through historical analysis, one could attempt to construct models of the decision-making strategies of past great scientists. The models might, given the same data base, duplicate the discoveries or evaluations of these scientists. Modeling research of this type may seem whimsical, but a program has already been constructed that successfully generated various scientific laws (Langley 1979). The modeling research could address several questions: How complex would models need to be to duplicate the great discoveries or judgments? Would analysis uncover any consistencies in the underlying judgment strategies used by scientists across different fields? What differences in the form or complexity of judgment strategies might exist among the great, the near great, and the average scientist?

Second, by utilizing the longitudinal information provided by historical study, one could look at the relationship between the success of theories and cognitive variables. A theory's success could be gauged by selected criteria, such as the period of time the theory was widely accepted or the number of studies it generated. (See p. 142 for a discussion of the historical evaluation of theories.) One could compare the cognitive strategies or frameworks underlying successful and unsuccessful theories. Would only unsuccessful theories be characterized by instances in which the theoretician failed to adhere to certain normative judgment guidelines? Would more successful theories be characterized by relatively simpler or more complex underlying cognitive frameworks? Would expansions in the underlying complexity of theories be associated with greater or lesser subsequent success? One could also attempt to identify factors that consistently differentiate successful and unsuccessful theories. These factors might pertain to the theory's "performance," such as its past success, or to its structural aspects, such as its parsimony or its underlying cognitive framework or reasoning strategies. If such factors could be identified, they might be used to construct actuarial programs for theory evaluation. Then, as described in chapter 4, one could use historical data to compare actuarial versus clinical methods of theory evaluation. Further, one might study which of these factors scientists identified when evaluating theories and the extent to which their judgments approached decision strategies that maximally discriminated successful and unsuccessful theories. Analysis of the match between actual and optimal decision rules could also be conducted with contemporary scientists.

NATURAL OBSERVATION

As applied to the study of science, natural observation strategies involve the observation of scientists in their normal work setting. Naturalistic studies of scientists generate rich information, which may account for their increasing popularity. There is no reason why such studies cannot include observation of certain judgment phenomena. One could observe the frequency and specific conditions under which confirmatory and disconfirmatory strategies are implemented. One could also observe the extent to which specific judgments adhered to normative decision rules and examine the relationship between adherence and progress on specific problems. In investigating the frequency and impact of violations of assumed normative judgment rules, such as violations based on belief in the "law of small numbers," might one find situations in which violation led to progress and thus was apparently not contraindicated? Another area for study is the ongoing problem solving of laboratory scientists. For example, one could attempt to analyze the complexity of their problem-solving strategies. In their normal work setting, would researchers exhibit proficiency with problem-solving strategies requiring cognitive complexity beyond the level assumed possible given the performance of subjects in judgment studies? In conducting the investigation, one could not accept self-reports uncritically. Individuals often claim to use very complex strategies, but experimental investigations have thus far not supported their claims. Some independent method for demonstrating complexity, other than self-report, would be required.

COMPARATIVE-ANALOGUE STRATEGIES

Comparative-analogue strategies encompass the study of lay people performing tasks holding positive analogy to tasks performed by scientists. These studies are comparative in that the subjects are not the population of interest, but they are similar to this population. Through the study of subjects that can be compared with the population of interest, one hopes to gain insight into the population and a knowledge base for generating more educated and productive hypotheses than would otherwise be possible. These studies are analogues in that the tasks that are used are related to, or hold positive analogy to, the tasks performed by scientists.

The comparative-analogue strategy can serve a number of purposes, but it is perhaps most useful when scientists are not available for study. This strategy holds promise for the investigation of major questions about scientific judgment. For example, one could use it to compare the capacity of individuals versus groups to perform complex problem-solving tasks, to generate complex models, or to judge the validity of proposed problem solutions that vary in degree of complexity. Would one be able to specify circumstances in which the cognitive capacity of the group is greater than the cognitive capacity of individuals? Some attempts have been made to study these questions. Gruber (conversation with the author, 1979) compared the ability of individuals versus groups to correctly uncover the properties of certain objects when limited information was available, a task analogous to scientific discovery. Mynatt, Doherty, and Tweney (1977, 1978) and others (see chap. 3) have studied the use of confimatory and disconfirmatory strategies in ongoing problem solving. One would hope that the studies would provide insight into the conditions in which one or the other strategy is preferable.

It is also possible to study the processes by which individuals form illusory beliefs about relationships or correlations between variables. Similarly, one can study the impact of new information on existent illusory beliefs—such as the conditions that encourage the maintenance of illusory beliefs in the face of disconfirming evidence—and the conditions that lead to changes in these beliefs. Would some analogue to "anomalous events" create change in belief, and what specific form or pattern of anomalous events would be required? Although the Chapmans (1967, 1969) and Golding and Rorer (1972) have investigated illusory beliefs, they were interested in the general manifestations of such beliefs, not in their specific occurrence in scientific or scientificlike settings. Researchers have rarely attempted to provide close analogies to scientific activities, which might offer further elucidation of the formation, maintenance, and alteration of illusory beliefs as manifested in science.

EXPERIMENTAL INVESTIGATIONS WITH SCIENTISTS

Research can be conducted in which scientists serve as subjects and variables are systematically manipulated and controlled. The virtual neglect of this powerful strategy, perhaps the most serious methodological oversight besetting the study of science, has precluded access to data that might advance our understanding of central questions. Among the possible studies is an analysis of the review process, as described previously. One would obtain two groups of scientists with differing positions on an issue and then have them evaluate the methodological adequacy of sets of studies confirming and disconfirming their respective positions. Another study, or more realistically a research program, could address the formation of hypotheses following exposure to information and the conditions under which these hypotheses are maintained or altered.

An ambitious but potentially valuable research program could address the relationship between exposure to information pertaining to a theory and the evaluation of that theory. One could simulate data or use hypothetical information to ensure control and generation of an adequate pool of experimental conditions. Variables could be manipulated to assess their impact on theory evaluation. As an enlarged data base was provided, would this additional base increase confidence in theory choice? Would confidence increase, even when the information was ambiguous and provided no additional support for the theory? How well could additional data be utilized when they did contain useful information? Would the previous finding—that additional information often does not increase (and can even decrease) judgment accuracy—generalize to the evaluation of scientific theories?

Experimental investigation of theory evaluation could also be directed towards other phenomena. After providing information about a theory and having scientists evaluate it, one could try to construct models to duplicate their judgments. Evaluations could be made of varying forms of information, ranging from raw data to theoretical postulates. The models of the scientists' judgments might provide insight into their underlying reasoning strategies. Mindful of the advantages that are often obtained through bootstrapping or the modeling judges, one might examine whether application of these models to other information was feasible or useful. Another line of investigation entails the systematic manipulation of evaluative criteria that are not data based and subsequent assessment of the impact of the manipulation on theory evaluation. One could experimentally manipulate the theory's logical consistency, parsimony, or congruency with the scientist's prior beliefs. Again, hypothetical theories could be used, increasing experimental control and the possibility of generating a sufficient pool of experimental conditions.

Many other experiments could be conducted to study theory evaluation or other aspects of scientific judgment. Researchers could compare individual and group judgment: Under what conditions and for which types of judgments would the group outperform the individual? Could instances be generated in which the group transcended the individual's upper limits in cognitive complexity? Efforts could also be directed towards detailed experimental analysis of the process of scientific problem solving. Investigators might compare the performance of elite and nonelite scientists on common scientific judgment tasks, or they might study the manner in which experience or expertise shapes judgment. For example, would repeated exposure to certain phenomena actually diminish the scientist's ability to "see" or comprehend alternative interpretations or explanations of these phenomena—would it increase closed-mindedness?

OBSTACLES TO RESEARCH

Many of these research questions and issues pose difficult problems for investigators. Numerous conceptual issues, such as the development of satisfactory criteria for normative judgments, are anything but straightforward, requiring an awareness of ongoing debates within the philosophy of science. Numerous practical problems also exist. For example, scientists must permit the investigator to scrutinize possible limitations in their reasoning. Further, although cognitive psychology offers certain meaningful insights and leads into scientific reasoning, with the study of human judgment limitations providing the most notable example, the field is still immature, and there are enough in-house problems to engage the attention of cognitive psychologists for a long time.

These problems restrict the contributions that psychologists can make to the study of science, but there is no doubt that we can improve upon the intuitive and often naive assumptions about human cognition that are so common to writings on science. Research in the psychology of science requires unusual ingenuity and breadth, but it is not impossible and has much to offer. It can contribute to our understanding of science and of human cognition in general by helping to outline the processes by which knowledge is acquired and the boundaries of human ability.

Notes

NOTES

Chapter 1 • Basic Assumptions and the Psychology of Science

1. From *Motiv de Forschens*. Quotation appears in Holton (1978, 377-78).

2. Hayek (1952) does not contend that the new sensory order necessarily represents a step towards a more "real" depiction of the world. His underlying reasoning is in large part original and defies any simple labels. The essentials of Hayek's views on the sensory order, which have been developed over many years and provide fundamental insights, are well represented in his 1952 book.

3. From *Induktion und deduktion in der physik*. Quotation appears in Holton (1978, 99).

4. Spencer Brown (1972), Weimer (1979), and others (e.g., Kuhn 1974; Polanyi 1966) might object to these comments about the role of theories in guiding scientific activities, especially if one is referring to scientific perceiving and the explicit content of formulated theories. Rather, they would argue that scientific perceiving is more like acquiring a (conceptual) language and that mastery of this language is achieved through practice and not by learning its explicit rules. The ability to perceive lies in the capacity to *use* the theory, to internalize it and apply its perceptual "vision," not in the acquisition of its content. Although this issue is relevant to many debates about science, it has little bearing on arguments made here about the complexity of factual perceiving.

5. To avoid digression in the body of the text, I have bypassed several considerations relevant to the possible interrelationship between strategy and world view. For example, the characteristics of this interrelationship likely vary in accordance with a series of variables, such as the maturity of the scientific field, the source from which the strategy originated (i.e., whether it emerged from within the field or was adopted from another field), and the level of congruence between the model of nature implicit in the strategy and the theoretical model of nature held by members of a community. In a mature field, which has used firmly established theoretical assumptions to develop its own specific strategies, the strategies are more likely to reflect theoretical views than to create them. In contrast, in an immature field that adopts strategies proven useful in other prestigious fields but that contain implicit assumptions varying widely from those prevalent in this immature field, the relationship is more likely to be reversed, with the strategy taking an important role in shaping theoretical

viewpoint. The latter conditions may apply to behavioral psychology. Additionally, regardless of these interrelationships, the particular forms of information produced by particular strategies do, in part, create a data base that has a potential influence upon the theoretical viewpoint. In general, little is known about the relationships between strategy and theoretical world view; work on this topic might provide useful insights.

6. It should be recognized, particularly when referring to mental processes that have become automatic and can be performed with little or no conscious effort, that the boundaries between perception and cognition are considerably blurred. In substantial part, the two processes are mutually dependent, and in many cases it is not even clear whether a mental operation is primarily perceptual or cognitive (e.g., "thinking" in visual images). In the study of scientific "seeing," and in the study of human mentation in general, we often cannot do much better than attempt to be clear about the type of activity to which we believe we are referring. We are far from developing methods for making exact distinctions or from a knowledge base that will allow precise description of the interactions between the two, especially when the subject matter is everyday scientific problem solving.

7. The quotation within the Holton quotation comes from an article by Weinberg (1974), which Holton uses extensively to provide examples of themata.

8. Conceptualizing social input as forms of information or as variables that influence access or exposure to information may be a starting point in explaining the mechanisms by which social or external factors shape belief, a problem that has been troublesome for the sociology of science. This conceptualization would lead one to argue that social factors have an impact upon belief through the information they provide or the manner in which they influence access or exposure to information. The full range of social factors can be seen as sharing this common property. In this simple model, social factors have an impact upon information, which in turn influences belief.

Chapter 2 • *Human Judgment Abilities*

1. There is a possible exception to this statement. Multivariate methods may capitalize on chance (nonrepeating) relationships between the predictive data and the criterion and, therefore, might result in a higher correlation between prediction and outcome. Hypothetically, the achieved level of correlation could even exceed that obtained if judges or formulae fully utilized the predictive potential of nonchance relationships. Any such increment based on chance relationships, however, will not generalize to new samples.

2. Although Sawyer's use of the term *mechanical* seems nonproblematic when addressing data collection, it does not necessarily retain Meehl's conceptualization of the actuarial method when applied to data combination. Upon this one point, Sawyer's article may create confusion rather than clarity. Methods for combining information can be mechanical and at the same time *not* actuarial. For example, one can program a computer to follow clinically derived decision procedures, a method that is obviously mechanical but not actuarial. Sawyer would probably have been better off sacrificing parsimony and using the clinical-mechanical distinction for data collection and retaining the clinical-actuarial distinction for data combination. (For further discussion see Marks, Seeman, and Haller [1974].) In practice, Sawyer's classification of methods for combining information generally follows the clinical-actuarial distinction. Therefore, in the following discussion of Sawyer's review, one can mentally substitute *clinical-actuarial* for *clinical-mechanical* when reference is made to methods for combining information.

3. Clinically collected data or impressions would need to be coded into a format, such as trait ratings, that could be utilized by actuarial methods. For the purpose of providing a fair test, both judge and formula would then be granted access to these coded ratings only. If the clinician had direct access to the patient, this would almost certainly provide information

that was not available for actuarial judgment, given the likelihood that all information obtained through direct contact could not be captured by the coding. This leaves open the possibility that information gathered through direct patient contact but not captured by the coded ratings would be unavailable to the clinician, deriving him or her of data critical to accurate clinical prediction. As Holt claimed, this might place the clinician at a relative disadvantage, compared with the normal conditions under which clinical predictions are made. One can, however, compare conditions in which clinicians are provided with all of the information available in everyday practice with conditions in which the actuarial method does not have access to all of this information, such as clinical impressions that cannot be coded. This is not a fair test of Meehl's specific hypothesis, but it does address Holt's contention and is one of the possible legitimate comparisons between clinical and actuarial judgment.

4. For the sake of clarity, Sawyer's other two conditions are not described here, although one of these, "clinical synthesis," is described later in the text. The other condition, "mechanical synthesis," is not relevant to the current discussion, and the findings bearing on this condition in no way conflict with any of the findings described in this chapter. For a description of mechanical synthesis, see Sawyer's (1966) article.

Chapter 3 • *Factors Underlying the Judgment Findings*

1. This unpublished study is described in Slovic (1976).

2. Meehl has since changed his position. During conversation with the author in December 1981, he expressed doubts about the human capacity to process cues in a configural manner and the potential advantages this provides the clinician over the actuary.

Chapter 4 • *Human Judgment and Science*

1. This is not to argue that all scientists in all fields make the same judgment errors. Scientific procedures and methods offer protection against certain judgment errors, and fields surely vary in the extent and types of protection offered. As should become clear during the reading of this chapter, although the broad issue of whether judgment findings do or do not generalize to science is primary (for if one believes they do not there is no problem to study), this issue does not seem to be in doubt. Therefore, it would be more productive to ask and investigate more refined questions: How do types and manifestations of judgment errors vary across certain fields; to what extent, and under what conditions, does mathematization offer safeguards against certain judgment errors; and what is the relationship between the type of judgments most commonly required within certain fields and the conventional procedures and methods utilized within these fields?

2. Of course, such results may have been obtained but not published because of a "failure" to achieve significant results.

3. The reasoning strategies that are ideal in general, and ideal for scientists in particular, are obviously far from certain or generally agreed upon. We actually have very little idea about what reasoning strategies scientists ought to apply, and such "oughts" may vary depending on the goals one wishes to attain. Therefore, when discussing generalization to science, one must be cautious and tentative in labeling judgment practices as violations of normative guidelines. Related difficulties exist when one attempts to label a scientific judgment as erroneous. If we can never know with certainty when we have identified truth—if there are no certain means for evaluating the veridicality of scientific constructs—then we can never be sure whether scientific judgments are erroneous or not. But this is not license for an "anything goes" attitude. Optimal reasoning strategies and veridicality (versus judgment error) can at least serve as proposed regulative ideals that we can hope to approach and develop. More will be said about these issues in chapter 5.

4. See the Eisenberg study as reported in Wender (1971, n. 1).

5. Methods of meta-analysis proposed and developed by such researchers as Glass, McGaw, and Smith (1982) may be a start in this direction.

6. We need to clarify this misleadingly simple statement about "nonrandom changes." The types of nonrandom differences or alterations varies depending upon whether one is referring to a category of a phenomenon or specific instances of that category. When one refers to a category, whether it represents a hypothetical construct or a physical entity, nonrandom alterations represent differences that exist among different instances of the category. For example, if the category is a hypothetical construct, such as aggressive behavior, two behaviors may differ in certain respects and still belong to the same category. The differences may be nonrandom in the sense that they do not necessarily represent disordered or nonsystematic variations. The level of difference that exists varies widely across the subject matter of different scientific fields, although it also varies across the categories found within particular scientific fields.

When one refers to specific or separate instances of a category, such as when a biologist refers to separate, specialized cells (e.g., Schwann cells), nonrandom variation refers to something else. The physical entities we study, even atomic ones, are dynamic systems that constantly undergo changes or perturbations; but through homeostatic mechanisms they often regain a dynamic state of balance and thereby maintain the same identity over time. For example, a Schwann cell is constantly undergoing changes while still maintaining its status as a Schwann cell. Although these perturbations may represent a reaction to environmental events, which themselves may be random, the variations the entity exhibits are a product of highly systematized and nonrandom internal operations. In this sense, the changes or differences in form that the entity undergoes do not represent random or disordered processes. Therefore, regardless of whether one refers to categories or specific instances within categories, nonrandom differences or variations are an inherent quality of the subject matter under consideration, although the specific form of variation may differ depending upon the subject matter to which one refers.

7. It may seem contradictory to state that a possible manifestation of cognitive limitations—bad judgment habits—are not that difficult to reverse but that the effects of cognitive limitations are. This is not actually a contradiction. Bad judgment habits do not always result from cognitive limitations, and, when they do not, they can usually be reversed. When bad habits do arise from cognitive limitations, their effects may still be mitigated, but this only provides awareness of possible erroneous judgments stemming from these habits and does not in any way reveal what the correct solution to the problem might be (as I will describe later in this chapter). Further, bad judgment habits are only one of many possible manifestations of cognitive limitations (assuming that bad habits are a manifestation of cognitive limitations), all of which involve an incapacity to solve a problem adequately or to make progress towards constructing satisfactory solutions.

8. Luria (1963) has actually designed remediation programs to improve these patients' capacity to comprehend complex grammatical relationships. The programs are designed, however, to teach patients new ways of approaching complex grammatical material, using intact brain areas that retain the capacity to manage more complicated information. These approaches, then, depend on currently available abilities; they do not provide generally increased abilities to manage complex information. Rather, they permit *generalization* of these current capacities to a content area (complex grammatical relationships) in which they were not previously applied.

9. There are, however, indirect ways to increase an individual's capacity to comprehend complex interrelations. Two possible ways are described in the section below entitled "Transcending Cognitive Limitations."

10. From a letter to C. Lanczos as quoted in Holton (1978, 144).

11. This may explain, in part, why the public often fails to comprehend scientific explanations. Two significant mechanisms for reducing cognitive strain — overlearning the relevant material and gaining familiarity with a system for representing or expressing data that condenses information — are not available. Therefore, although the cognitive capacities of the lay person may often equal that of the scientist, he or she may face a task requiring far greater cognitive complexity when attempting to understand what the scientist understands.

12. This position, of course, holds parallels to the evolutionary epistemology of Campbell (1974) and Popper (1972), although neither proponent directly addresses the possibility that limitations in the capacity to simultaneously hold information in mind may be a critical factor underlying the mechanisms through which human knowledge develops.

13. The question of whether scientific constructions arising from simplifying strategies reflect mind or external reality is probably too global to be useful. It is perhaps more useful to question the extent to which such constructions reflect internal mental structure or function, which may or may not be congruent with external reality; the extent to which they reflect external reality; and the manner in which and the mechanisms through which internal structure and function and external information interact. The essential task, then, becomes to learn how the mind and the external world interact in the process of constructing external reality, and one need not assume that the mind's contributions distort external reality. Rather, such contributions can be seen as a means for interpretation, which may lead to good approximations of external reality, but approximations nonetheless.

At the same time, human thinking has characteristics and limitations that give form to reality and set restrictions on what can be seen. Relative to the issue of cognitive limitations, these limitations probably set boundaries on the reality constructions that are possible and also create associated tendencies to prefer or use certain strategies for achieving the reality constructions that fall within these boundaries. An example is the strong tendency to utilize simple reasoning strategies when attempting to solve problems. For the scientist, this may result in erroneous or distorted constructions when nature will not yield to simpler strategies. Given the above considerations, no unqualified answer can be given to the question of whether simplifying strategies result in good approximations or distortions of reality. Some degree of both is always present, and our task is to determine how much of each is present and, when possible, to identify and reduce distortion.

14. This topic is treated in my work in progress, sections of which are available by request.

15. My reading of the current state of systems approaches is that the most successful efforts have relied heavily on decision aids, such as computer methods and (related to this) sophisticated mathematical procedures. Verbal approaches that do not incorporate these aids have been less successful and are perhaps most accurately described as frameworks that permit greater understanding, rather than as theories that are useful in formulating specific predictions. The current popularity of these verbal approaches may be a by-product of the successes of more rigorous approaches, which have established general scientific credibility for systems theories. Their potential for incorporating decision aids and mathematical procedures that permit adequate consideration of the complex interrelations they attempt to describe may ultimately determine the success of such verbal approaches.

16. Even if, for some unforeseen reason, we are unable to increase the computer's capacity to detect complex interrelationships, the capacity that has thus far been developed could aid in such judgments. For example, one could model the reasoning strategies of leading scientists and then apply these models to new sets of data.

Chapter 5 • *Reflections, Extensions, and Elaborations of the Argument*

1. These issues are treated further in my work in process, sections of which are available by request.

2. Complex *perceptual* acts are doubtless possible, but the capacity to perform them is not sufficient to perform the complex *cognitive* judgments discussed in this book.

3. As those familiar with the work of Piaget will recognize, these statements about necessary conditions for the development of thought represent applications of Piagetian concepts, particularly those pertaining to "accommodation" and "assimilation" (for an overview of Piaget's work, see Flavell [1963]).

4. In using Kuhn's concepts of normal and revolutionary science to characterize the two types of "steps beyond," I do not mean to imply that the latter—revolutions in thought —need to involve entire paradigms or all-encompassing ideas. What I would consider reorganization or qualitative changes in thought can occur on a much narrower scale, as within small pockets of theoretical systems.

5. Beliefs about how simple or complex principles must be in order to explain the world tend to, but do not *necessarily*, correspond with beliefs about the complexity present in the world or the complexity of thought needed to uncover these principles. For example, one can believe that the world is very complex but that simple principles can provide an excellent approximation; or that the world is basically simple but that very complex reasoning abilities will be necessary to discover underlying principles.

References

REFERENCES

Allport, D. A., B. Antonis and P. Reynolds. 1972. On the division of attention. *Quarterly Journal of Experimental Psychology* 24:225-35.

Anderson, N. H. 1972. *Information integration theory: A brief survey*. Center for Human Information Processing Technical Report, La Jolla. University of California, San Diego.

_____.1974. *The problem of change-of-meaning*. Center for Human Information Processing Technical Report, La Jolla. University of California, San Diego.

Anderson, N. H. and A. Jacobson. 1965. Effect of stimulus inconsistency and discounting instructions in personality impression formation. *Journal of Personality and Social Psychology* 2:531-39.

Arden, B. W., ed. 1980. *What can be automated?* Cambridge, Mass.: MIT Press.

Asch, S. 1946. Forming impressions of personality. *Journal of Abnormal and Social Psychology* 41:258-90.

Baker, W. O., W. S. Brown, M. V. Mathews, S. P. Morgan, H. O. Pollak, R. C. Prim and S. Sternberg. 1977. Computers and research. *Science* 195:1134-39.

Bandura, A. 1974. Behavior theory and the models of man. *American Psychologist* 29:859-69.

Bean, L. H. 1979. The logical reasoning abilities of scientists: An empirical investigation. Master's thesis, Ohio State University.

Bem, D. J. 1972. Self-perception theory. In *Advances in experimental social psychology*, vol. 6, edited by L. Berkowitz. New York: Academic Press.

Bertalanffy, L. von. 1968. *General systems theory: Foundations, developments, applications*. New York: George Braziller.

Beveridge, W. I. B. 1957. *The art of scientific investigation*. Rev. ed. New York: Random House.

Birnbaum, M. H. 1973. The devil rides again: Correlation as an index of fit. *Psychological Bulletin* 79:239-42.

Böhme, G. 1977. Models for the development of science. In *Science, technology and society*, edited by I. Spiegel-Rösing and D. de Solla Price. London: Sage Publications.

Bowers, K. S. 1973. Situationalism in psychology: An analysis and critique. *Psychological Review* 80:307-36.

Brehmer, B. 1969. Cognitive dependence on additive and configural cue-criterion relations. *American Journal of Psychology* 82:490-503.

Bunge, M. 1967. *Scientific research.* vols. 1 and 2. New York: Springer-Verlag.

Campbell, D. T. 1974. Evolutionary epistemology. In *The philosophy of Karl Popper*, edited by P. A. Schilpp. Vol. 14. *The library of living philosophers.* La Salle, Ill.: Open Court Publishing Co.

_____. 1978. Comments on the fragmented history of the psychology of science. Paper presented at the Society for the Social Study of Science, Bloomington, Ind. November.

Campbell, D. T. and J. C. Stanley. 1963. *Experimental and quasi-experimental designs for research.* Chicago: Rand McNally & Company.

Capute, A. J., E. F. L. Niedermeyer and F. Richardson. 1968. The electroencephalogram in children with minimal cerebral dysfunction. *Pediatrics* 41:1104-14.

Chapman, L. J. and J. P. Chapman. 1967. Genesis of popular but erroneous psychodiagnostic observations. *Journal of Abnormal Psychology* 72:193-204.

_____. 1969. Illusory correlation as an obstacle to the use of valid psychodiagnostic signs. *Journal of Abnormal Psychology* 74:271-80.

Crane, D. 1972. *Invisible colleges.* Chicago: University of Chicago Press.

Cronbach, L. J. 1957. The two disciplines of scientific psychology. *American Psychologist* 12:671-84.

Cronbach, L. J. and P. E. Meehl. 1955. Construct validity in psychological tests. *Psychological Bulletin* 52:281-302.

Dawes, R. M. 1964. Social selection based on multidimensional criteria. *Journal of Abnormal and Social Psychology* 68:104-9.

_____. 1971. A case study of graduate admissions: Application of three principles of human decision making. *American Psychologist* 26:180-88.

_____. 1976. Shallow psychology. In *Cognition and social behavior*, edited by J. S. Carroll and J. W. Payne. Hillsdale, N. J.:Lawrence Erlbaum Associates.

Dawes, R. M. and B. Corrigan. 1974. Linear models in decision making. *Psychological Bulletin* 81:95-106.

Descartes, R. 1912. *Discourse on method*, translated by J. Veitch. Originally published in 1637. London: Everyman Library.

Deutsch, M. 1959. Evidence and inference in nuclear research. In *Evidence and inference*, edited by D. Lerner. New York: Free Press of Glencoe.

Dudycha, A. L. and J. C. Naylor. 1966. The effects of variations in the cue R matrix upon the obtained policy equation of judges. *Educational and Psychological Measurement* 26:583-603.

Durkheim, E. 1915. *The elementary forms of the religious life.* London: George Allen and Unwin.

Earle, T. C. 1970. Task learning, interpersonal learning, and cognitive complexity. *Oregon Research Institute Research Bulletin*, vol. 10, no. 2.

Eccles, J. C. 1977. *The understanding of the brain.* New York: McGraw-Hill Book Company.

Einhorn, H. J. 1971. Use of nonlinear, noncompensatory models as a function of task and amount of information. *Organizational Behavior and Human Performance* 6:1-27.

_____. 1972. Expert measurement and mechanical combination. *Organizational Behavior and Human Performance* 7:86-106.

Feingold, B. F. 1975. *Why your child is hyperactive.* New York: Random House.

Feyerabend, P. 1979. Dialogue on method. In *The structure and development of science*, edited by G. Radnitzky and G. Anderrson. Dordrecht, Holland: D. Reidel Publishing Company.

Findler, N. V., ed. 1979. *Associative networks — the representation and use of knowledge in computers.* New York: Academic Press.

Fischhoff, B., P. Slovic and S. Lichtenstein. 1977. Knowing with certainty: The appropriateness of extreme confidence. *Journal of Experimental Psychology: Human Perception and Performance* 3:552-64.

Flavell, J. H. 1963. *The developmental psychology of Jean Piaget.* New York: D. Van Nostrand Company.

Garfield, S. L. and A. E. Bergin, eds. 1978. *Handbook of psychotherapy and behavior change: An empirical analysis.* 2d ed. New York: John Wiley.

Gibson, E. J. 1969. *Principles of perceptual learning and development.* New York: Appleton-Century-Crofts.

Glass, G. V., B. McGaw and M. L. Smith. 1982. *Meta-analysis in social research.* Beverly Hills, Calif.: Sage Publications.

Goldberg, L. R. 1959. The effectiveness of clinicians' judgments: The diagnosis of organic brain damage from the Bender-Gestalt Test. *Journal of Consulting Psychology* 23: 25-33.

_____. 1965. Diagnosticians versus diagnostic signs: The diagnosis of psychosis versus neurosis from the MMPI. *Psychological Monographs* 79 (9, Whole No. 602).

_____. 1968a. Seer over sign: The first "good" example? *Journal of Experimental Research in Personality* 3:168-71.

_____. 1968b. Simple models or simple processes? Some research on clinical judgments. *American Psychologist* 23, 483-96.

_____. 1970. Man versus model of man: A rationale, plus some evidence, for a method of improving on clinical inferences. *Psychological Bulletin* 73:422-32.

Golden, M. 1964. Some effects of combining psychological tests on clinical inferences. *Journal of Consulting Psychology* 28:440-46.

Golding, S. L. and L. G. Rorer. 1972. Illusory correlation and subjective judgment. *Journal of Abnormal Psychology* 80:249-60.

Gollob, H. F. 1968. Impression formation and word combination in sentences. *Journal of Personality and Social Psychology* 10:341-53.

Gottesman, I. I. and J. Shields. 1972. *Schizophrenia and genetics: A twin study vantage point.* New York: Academic Press.

Hammond, K. R. and D. A. Summers. 1965. Cognitive dependence on linear and nonlinear cues. *Psychological Review* 72:215-24.

Hanson, N. R. 1958. *Patterns of discovery.* Cambridge: Cambridge University Press.

Harré, R. 1972. *The philosophies of science.* London: Oxford University Press.

Hayek, F. A. 1952. *The sensory order.* Chicago: University of Chicago Press.

_____. 1955. *The counter-revolution of science.* New York: Free Press of Glencoe.

_____. 1969. The primacy of the abstract. In *Beyond reductionism,* edited by A. Koestler and J. R. Smythies. New York: Macmillan.

Heisenberg, W. 1971. *Physics and beyond.* New York: Harper & Row.

_____. 1973. Development of concepts in the history of quantum theory. Lecture at Harvard University, May. Reprinted in *American Journal of Physics* 43 (1975):389-94.

Hislop, M. W. and L. R. Brooks. 1968. Suppression of concept learning by verbal rules. Technical Report No. 28, Department of Psychology, McMasters University, December.

Hoepfl, R. T. and G. P. Huber. 1970. A study of self-explicated utility models. *Behavioral Science* 15:408-14.

Hoffman, P. J. 1960. The paramorphic representation of clinical judgment. *Psychological Bulletin* 57:116-31.

Hoffman, P. J., P. Slovic and L. G. Rorer. 1968. An analysis-of-variance model for the assessment of configural cue utilization in clinical judgment. *Psychological Bulletin* 69:338-49.

Holt, R. R. 1958. Clinical and statistical prediction: A reformulation and some new data. *Journal of Abnormal and Social Psychology* 56:1-12.

Holton, G. 1973. *Thematic origins of scientific thought: Kepler to Einstein.* Cambridge, Mass.: Harvard University Press.

_____. 1978. *The scientific imagination: Case studies.* Cambridge: Cambridge University Press.

Horvath, F. S. 1977. The effect of selected variables on interpretation of polygraph records. *Journal of Applied Psychology* 62:127-36.

Huber, G. P., V. K. Sahney and D. L. Ford. 1969. A study of subjective evaluation models. *Behavioral Science* 14:483-89.

Hughes, J. R. 1976. Biochemical and electroencephalographic correlates of learning disabilities. In *The neuropsychology of learning disorders*, edited by R. M. Knights and D. J. Bakker. Baltimore: University Park Press.

Hull, C. L. 1943. *Principles of behavior.* New York: Appleton-Century-Crofts. 1952.

_____. 1952. *A behavior system.* New Haven, Conn.: Yale University Press.

Jones, E. E. and G. Goethals. 1972. Order effects in impression formation: Attribution context and the nature of the entity. In *Attribution: Perceiving the causes of behavior*, edited by E. E. Jones et al. Morristown, N. J.: General Learning Press.

Kahneman, D. 1973. *Attention and effort.* Englewood Cliffs, N. J.: Prentice-Hall, Inc.

Kahneman, D. and A. Tversky. 1972. Subjective probability: A judgment of representativeness. *Cognitive Psychology* 3:430-51.

_____. 1973. On the psychology of prediction. *Psychological Review* 80:237-51.

Kelley, H. H. 1967. Attribution theory in social psychology. In *Nebraska Symposium on Motivation,* vol. 15, edited by D. Levine. Lincoln: University of Nebraska Press.

_____. 1972. Attribution in social interaction. In *Attribution: Perceiving the causes of behavior,* edited by E. E Jones et al. Morristown, N. J.: General Learning Press.

_____. 1973. The process of causal attribution. *American Psychologist* 28:107-28.

Kort, F. 1968. A nonlinear model for the analysis of judicial decisions. *American Political Science Review* 62:546-55.

Kostlan, A. 1954. A method for the empirical study of psychodiagnosis. *Journal of Consulting Psychology* 18:83-88.

Kubie, L. S. 1954. Some unsolved problems of the scientific career, II. *American Scientist* 42:104-12.

Kuhn, D., J. Langer, L. Kohlberg and N. Haan. 1977. The development of formal operations in logical and moral judgment. *Genetic Psychology Monographs* 95:97-188.

Kuhn, T. S. 1970. *The structure of scientific revolutions.* 2d ed. Chicago: University of Chicago Press.

_____. 1974. Second thoughts on paradigms. In *The structure of scientific theories,* edited by F. Suppe. Urbana: University of Illinois Press.

Lakatos, I. 1968. Criticism and the methodology of scientific research programmes. *Proceedings of the Aristotelian Society* 69:149-86.

_____. 1970. Falsification and the methodology of scientific research programmes. In *Criticism and the growth of knowledge,* edited by I. Lakatos and A. Musgrave. Cambridge: Cambridge University Press.

_____. 1972. History of science and its rational reconstructions. In *Boston studies in the philosophy of science,* vol. 8, edited by R. Buck and R. S. Cohen. New York: Humanities Press.

Lampel, A. K. and N. H. Anderson. 1968. Combining visual and verbal information in an impression-formation task. *Journal of Personality and Social Psychology* 9:1-6.

Langley, P. 1979. Rediscovering physics with BACON. *Proceedings of the Sixth International Joint Conference on Artificial Intelligence*, 505-7.

Latour, B. and S. Woolgar. 1979. *Laboratory life: The social construction of scientific facts.* Beverly Hills, Calif.: Sage Publications.

Levine, M. 1969. Neo-noncontinuity theory. In *The psychology of learning and motivation*, vol. 3, edited by G. H. Bower and J. T. Spence. New York: Academic Press.

Lewis, C. I. 1929. *Mind and the world order.* Cambridge, Mass.: Harvard University Press. Reprint. New York: Dover Publications, 1956.

Lord, C., L. Ross and M. R. Lepper. 1979. Biased assimilation and attitude polarization: The effects of prior theories on subsequently considered evidence. *Journal of Personality and Social Psychology* 37:2098-109.

Luria, A. R. 1963. *Restoration of function after brain injury.* Translated by B. Haigh. New York: Macmillan.

_____. [1962] 1980. *Higher cortical functions in man.* 2d ed. Translated by B. Haigh. New York: Basic Books.

Lykken, D. T. 1979. The detection of deception. *Psychological Bulletin* 86:47-53.

Lynch, M. 1981. *Art and artifact in laboratory science.* London: Routledge and Kegan Paul.

MacLeod, R. 1977. Changing perspectives in the social history of science. In *Science, technology and society*, edited by I. Spiegel-Rösing and D. de Solla Price. London: Sage Publications.

Mahoney, M. J. 1976. *Scientist as subject.* Cambridge, Mass.: Ballinger.

_____. 1977. Publication prejudices: An experimental study of confirmatory bias in the peer review system. *Cognitive Therapy and Research* 1:161-75.

Mahoney, M. J. and B. G. DeMonbreun. 1977. Psychology of the scientist: An analysis of problem-solving bias. *Cognitive Therapy and Research* 1:229-38.

Mahoney, M. J. and T. P. Kimper. 1976. From ethics to logic: A survey of scientists. In *Scientist as subject*, by M. J. Mahoney. Cambridge, Mass.: Ballinger.

Marx, M. H. and W. A. Hillix. 1973. *Systems and theories in psychology.* New York: McGraw-Hill Book Company.

Masterman, M. 1980. Braithwaite and Kuhn: Analogy-clusters within and without hypothetico-deductive systems in science. In *Science, belief and behavior*, edited by D. H. Mellor. Cambridge: Cambridge University Press.

McArthur, C. C. 1956a. Clinical versus actuarial prediction. *Proceedings of the 1955 Invitational Conference on Testing Problems.* Princeton, N. J.: Educational Testing Service.

_____. 1956b. The dynamic model. *Journal of Counseling Psychology* 3:168-71.

McArthur, L. A. 1972. The how and what of why: Some determinants and consequences of causal attribution. *Journal of Personality and Social Psychology* 22:171-93.

McCorduck, P. 1979. *Machines who think.* San Francisco: Freeman.

Meehl, P. E. 1954. *Clinical versus statistical prediction: A theoretical analysis and a review of the evidence.* Minneapolis: University of Minnesota Press.

_____. 1965. Seer over sign: The first good example. *Journal of Experimental Research in Personality* 1:27-32.

_____. 1978. Theoretical risks and tabular asterisks: Sir Karl, Sir Ronald, and the slow progress of soft psychology. *Journal of Consulting and Clinical Psychology* 46:806-34.

Meehl, P. E. and A. Rosen. 1955. Antecedent probability and the efficiency of psychometric signs, patterns, or cutting scores. *Psychological Bulletin* 52:194-216.

Mitroff, I. 1974. *The subjective side of science.* Amsterdam: Elsevier Scientific Publishing Co.

Mynatt, C. R., M. E. Doherty and R. D. Tweney. 1977. Confirmation bias in a simulated research environment: An experimental study of scientific inference. *Quarterly Journal
of Experimental Psychology* 29:85-95.

————. 1978. Consequences of confirmation and disconfirmation in a simulated research environment. *Quarterly Journal of Experimental Psychology* 30:395-406.

Newell, A. and H. A. Simon. 1972. *Human problem solving.* Englewood Cliffs, N. J.: Prentice-Hall.

Newton, I. 1931. *Opticks.* Originally published in 1704. London: Bell & Hyman.

Nisbett, R. E., E. Borgida and R. Crandall. 1976. Popular induction: Information is not
necessarily informative. In *Cognition and social behavior*, edited by J. S. Carroll and
J. W. Payne. Hillsdale, N. J.: Lawrence Erlbaum Associates.

Nisbett, R. E. and L. Ross. 1980. *Human inference: Strategies and shortcomings of social
judgment.* Englewood Cliffs, N. J.: Prentice-Hall, Inc.

Nisbett, R. E. and T. D. Wilson. 1977a. The halo effect: Evidence for unconscious alteration
of judgments. *Journal of Personality and Social Psychology* 35:250-56.

————. 1977b. Telling more than we can know: Verbal reports on mental processes.
Psychological Review 84:231-59.

Oskamp, S. 1962. How clinicians make decisons from the MMPI: An empirical study. Paper
presented at the American Psychological Association, St. Louis.

————. 1965. Overconfidence in case-study judgments. *Journal of Consulting Psychology* 29:261-65.

Payne, J. W. and M. L. Braunstein. 1971. Preferences among gambles with equal underlying
distributions. *Journal of Experimental Psychology* 87:13-18.

Petersén, I. and O. Eeg-Olofsson. 1971. The development of the electroencephalogram in
normal children from the age of 1 through 15 years—non-paroxysmal activity.
Neuropädiatrie 2:247-304.

Piaget, J. 1952. *The origins of intelligence in children.* Translated by M. Cook. New York:
International Universities Press.

————. 1970. *Genetic epistemology.* Translated by E. Duckworth. New York: Columbia University Press.

————. 1971. *Insights and illusions of philosophy.* Translated by W. Mays. New York:
World Publishing Company.

Polanyi, M. 1966. *The tacit dimension.* Garden City, N. Y.: Doubleday.

Popper, K. R. 1950. *The open society and its enemies.* Princeton, N. J.: Princeton University Press.

————. [1935] 1959. *The logic of scientific discovery.* New York: Harper.

————. 1972. *Objective knowledge: An evolutionary approach.* Oxford: Clarendon
Press.

————. [1962]1978. *Conjectures and refutations.* London: Routledge & Kegan Paul.

Posner, M. I. 1973. *Cognition: An introduction.* Glenview, Ill.: Scott, Foresman and Company.

Posner, M. I. and S. W. Keele. 1968. On the genesis of abstract ideas. *Journal of Experimental Psychology* 77:353-63.

Radnitzky, G. 1979. Justifying a theory vs. giving good reasons for preferring a theory. In
The structure and development of science, edited by G. Radnitzky and G. Andersson.
Boston: R. Reidel Publishing Company.

Restle, F. and E. R. Brown. 1970. Serial pattern learning. *Journal of Experimental Psychology* 83:120-25.

Roe, A. 1953. *The making of a scientist.* New York: Dodd, Mead.

Rorer, L. G., P. J. Hoffman, H. R. Dickman, and P. Slovic. 1967. Configural judgments revealed. *Proceedings of the 75th Annual Convention of the American Psychological
Association* 2:195-96.

Ross, L., M. R. Lepper and M. Hubbard. 1975. Perseverance in self-perception and social perception: Biased attributional processes in the debriefing paradigm. *Journal of Personality and Social Psychology* 32:880-92.

Samuel, A. L. 1967. Some studies in machine learning using the game of checkers, II—recent progress. *IBM Journal of Research and Development* 601-17.

Sawyer, J. 1966. Measurement *and* prediction, clinical *and* statistical. *Psychological Bulletin* 66:178-200.

Sellars, W. S. 1963. *Science, perception, and reality*. New York: Humanities Press.

Shaffer, L. H. 1974. Multiple attention in transcription. In *Attention and performance*, vol. 5, edited by P. M. A. Rabbitt. New York: Academic Press.

Sidowski, J. B. and N. H. Anderson. 1967. Judgments of city-occupation combinations. *Psychonomic Science* 7:279-80.

Simon, H. A. 1957. *Models of man: Social and rational*. New York: John Wiley.

_____. 1976. Scientific discovery and the psychology of problem solving. In *Models of discovery*, edited by H. A. Simon. Boston: D. Reidel Publishing Company.

_____. 1977. What computers mean for man and society. *Science* 195:1186-91.

_____. 1981. *The sciences of the artificial*. 2d ed. Cambridge, Mass.: MIT Press.

Sines, L. K. 1959. The relative contribution of four kinds of data to accuracy in personality assessment. *Journal of Consulting Psychology* 23:483-92.

Slovic, P. 1966. Cue-consistency and cue-utilization in judgments. *American Journal of Psychology* 79:427-34.

_____. 1969. Analyzing the expert judge: A descriptive study of a stockbroker's decision processes. *Journal of Applied Psychology* 53:255-63.

_____. 1976. Toward understanding and improving decisions. In *Science, technology and the modern navy: Thirtieth anniversary, 1946-1976*, edited by E. I. Salkovitz. Arlington, Va.: Office of Naval Research.

Slovic, P., and S. C. Lichtenstein. 1968a. The importance of variance preferences in gambling decisions. *Journal of Experimental Psychology* 78:646-54.

_____. 1968b. The relative importance of probabilities and payoffs in risk taking. *Journal of Experimental Psychology Monograph Supplement* 78, No. 3, part 2.

_____. 1971. Comparison of Bayesian and regression approaches to the study of information processing in judgment. *Organizational Behavior and Human Performance* 6:649-744.

Slovic, P. and D. MacPhillamy. 1974. Dimensional commensurability and cue utilization in comparative judgment. *Organizational Behavior and Human Performance* 11:172-94.

Snyder, M. 1978. Seek, and ye shall find: Testing hypotheses about other people. Paper presented at the Ontario Symposium on Personality and Social Psychology. London, Ontario, August.

Spencer Brown, G. 1972. *Laws of form*. New York: Julian Press.

Starr, B. J. and E. S. Katkin. 1969. The clinician as aberrant actuary: Illusory correlation and the Incomplete Sentences Blank. *Journal of Abnormal Psychology* 74:670-75.

Storms, M. D. 1973. Videotape and the attribution process: Reversing actors' and observers' points of view. *Journal of Personality and Social Psychology* 27:165-75.

Summers, D. A. 1967. Rule versus cue learning in multiple probability tasks. *Proceedings of the 75th Annual Convention of the American Psychological Association* 2:43-44.

Summers, D. A. and K. R. Hammond. 1966. Inference behavior in multiple-cue tasks involving both linear and nonlinear relations. *Journal of Experimental Psychology* 71:751-57.

Summers, D. A. and T. R. Stewart. 1968. Regression models of foreign policy judgments. *Proceedings of the 76th Annual Convention of the American Psychological Association* 3:195-96.

Summers, S. A., R. C. Summers and V. T. Karkau. 1969. Judgments based on different functional relationships between interacting cues and a criterion. *American Journal of Psychology* 82:203-11.

Taylor, S. E. and S. T. Fiske. 1975. Point of view and perceptions of causality. *Journal of Personality and Social Psychology* 32:439-45.

Turner, M. B. 1967. *Psychology and the philosophy of science.* New York: Appleton-Century-Crofts.

Tversky, A. 1969. Intransitivity of preferences. *Psychological Review* 76:31-48.

_____. 1977. Features of similarity. *Psychological Review* 84:327-52.

Tversky, A. and D. Kahneman. 1971. Belief in the law of small numbers. *Psychological Bulletin* 76:105-10.

_____. 1973. Availability: A heuristic for judging frequency and probability. *Cognitive Psychology* 5:207-32.

_____. 1974. Judgment under uncertainty: Heuristics and biases. *Science* 183:1124-31.

_____. 1978. Causal schemata in judgments under uncertainty. In *Progress in social psychology,* edited by M. Fishbein. Hillsdale, N. J.: Lawrence Erlbaum Associates.

Tweney, R. D., M. E. Doherty and C. R. Mynatt, eds. 1981. *On scientific thinking.* New York: Columbia University Press.

Vygotsky, L. S. 1962. *Thought and language.* Translated by E. Hanfmann and G. Vakar. Cambridge: MIT Press.

Ward, W. D. and H. M. Jenkins. 1965. The display of information and the judgment of contingency. *Canadian Journal of Psychology* 19:231-41.

Wason, P. and P. N. Johnson-Laird. 1972. *Psychology of reasoning: Structure and content.* Cambridge, Mass.: Harvard University Press.

Weimer, W. B. 1974a. The history of psychology and its retrieval from historiography: II. Some lessons for the methodology of scientific research. *Science Studies* 4:367-96.

_____. 1974b. The history of psychology and its retrieval from historiography: I. The problematic nature of history. *Science Studies* 4:235-58.

_____. 1977. Scientific inquiry, assessment, and logic: Comments on Bowers and Mahoney-DeMonbreun. *Cognitive Therapy and Research* 1:247-53.

_____. 1979. *Notes on the methodology of scientific research.* Hillsdale, N. J.: Lawrence Erlbaum Associates.

Weinberg, S. 1974. Unified theories of elementary-particle interaction. *Scientific American* 231 (1):50-59.

Weiner, I. B. 1970. *Psychological disturbance in adolescence.* New York: John Wiley.

Weizenbaum, J. 1976. *Computer power and human reason.* San Francisco: Freeman.

Wender, P. H. 1971. *Minimal brain dysfunction in children.* New York: John Wiley.

Wiggins, J. S. 1973. *Personality and prediction: Principles of personality assessment.* Reading, Mass.: Addison-Wesley Publishing Company.

Wiggins, N. and P. J. Hoffman. 1968. Three models of clinical judgment. *Journal of Abnormal Psychology* 73:70-77.

Wilson, T. D. and R. E. Nisbett. 1978. The accuracy of verbal reports about the effects of stimuli on evaluations and behavior. *Social Psychology* 41:118-31.

Winch, R. F. and D. M. More. 1956. Does TAT add information to interview? Statistical analysis of the increment. *Journal of Clinical Psychology* 12:316-21.

Yntema, D. B. and W. S. Torgerson. 1961. Man-computer cooperation in decisions requiring common sense. *IRE Transactions of the Professional Group on Human Factors in Electronics.* HFE 2 (1):20-26.

Zenzen, M. and S. Restivo. 1982. The mysterious morphology of immiscible liquids: A study of scientific practice. *Social Science Information* 21:447-73.

Index

David Faust earned his doctorate in psychology at Ohio University in 1980. He was a postdoctoral fellow and subsequently a faculty member in the Division of Child and Adolescent Psychiatry at the University of Minnesota. Faust is now director of psychology at Rhode Island Hospital and an assistant professor in the Brown University Medical School. He is the co-author of *Teaching and Moral Reasoning: Theory and Practice.*